Kiang The Theory of Fixed Point Classes

Kiang Tsai-han

The Theory of Fixed Point Classes

With 39 Figures

Springer-Verlag Berlin Heidelberg GmbH

Kiang Tsai-han

Department of Mathematics
Peking University
Beijing
The People's Republic of China

Mathematics Subject Classification (1980): 55 M 20

ISBN 978-3-642-68135-6 ISBN 978-3-642-68133-2 (eBook)
DOI 10.1007/978-3-642-68133-2

Library of Congress Cataloging-in-Publication Data
Chiang, Tse-han.
The theory of fixed point classes / Kiang Tsai-han p. cm.
"English version . . . of the . . . revised Chinese edition . . . In the English version, Appendix D and the well-known examples and proofs in the appendices A and B of the revised Chinese edition are omitted" - P. v.
Bibliography: p. Includes index.
ISBN 978-3-642-68135-6
1. Fixed point theory. I. Title. QA329.9.C49 1989 514-dc 19 89-5945 CIP

Originally published by Springer-Verlag Berlin Heidelberg and Science Press Beijing New York in 1989
Softcover reprint of the hardcover 1st edition 1989

Typesetting: Science Press, Beijing, The People's Republic of China
2141/3140-543210 Printed on acid-free paper

Preface to the English Version

Since the publication of the 1979 Chinese edition of the book *Theory of Fixed Point Classes,* my colleagues and I have used it as a major reference book in several seminars for beginning graduate students. When we began to revise it, our publishers proposed to translate it into English. We finally agreed that the author would translate the forthcoming revised Chinese edition. We explain here how the revised Chinese edition differs from the first Chinese edition and the English version from the revised.

The revised edition contains no new material except for a few further developments of some topics already treated in the first Chinese edition. A number of typographic and minor errors have been corrected, some new examples added, and numerous improvements on representation have been introduced.

In the English version, Appendix D and the well-known examples and proofs in the Appendices A and B of the revised Chinese edition are omitted, for they are easily accessible in the English literature.

The author wishes to express his gratitude to his colleagues and students for all their help in preparing the manuscripts of these three editions. For this English version in particular, gratitude is also extended to Prof. Jiang Boju for reviewing the manuscript, to Mr. Chang Huicuan for proofreading and compiling the index and the list of symbols, and to Springer-Verlag and to Science Press for English language assistance.

Peking University
August 1987

Kiang Tsai-han

Preface to the First Chinese Edition

In the vast ocean of mathematical literature, there are a great number of texts for university undergraduates and innumerable research papers in the mathematical journals. To the reader it is far more difficult to study the papers than to follow textbooks; there is a wide gap between the level of texts and that of papers. Thus in the literature there also exist numerous treatises or monographs, which serve to bridge this gap. For a beginning graduate, an appropriate treatise is of immense help.

In the last fifty years and more, some of the research papers on algebraic topology have given rise to a special topic called the theory of fixed point classes. So far as we know, there is as yet no undergraduate text which contains an account of the theory. The theory aims at determining the least number of fixed points of a self-mapping f of a topological space X, and is chiefly based on J. Nielsen's fundamental concept of separation of the fixed points into fixed point classes. This concept was first introduced only for the case that X is a torus, in a paper written in Danish (cf. [21]), and aroused vast interest when [32, I] appeared, in which X is an orientable closed surface. The space X in subsequent related papers has been generalized from orientable closed surface to closed surface, to finite polyhedron, to compact absolute neighborhood retract, and even to general topological space. Along with the development of the theory, applications have naturally also emerged in some other branches of mathematics.

This book is intended as an easy and introductory treatise on the theory of fixed point classes. We confine ourselves to the theory for a connected finite polyhedron, because it is the core of the whole theory. In the presentation, we try to proceed from the particular to the general, and from the concrete to the abstract; we start from examples, emphasize geometrical and physical explanations, and then give mathematical proofs. Not only will the geometrical meaning of the fundamental concepts be brought to the fore, but also some aspects of the historical development will be noted. In short, when we deliberated how to arrange the content of this book and how to present it, we chose to proceed at a leisurely pace and not to forget the geometrical background so that the reader would not be scared away after opening the book, and endeavour to pave the way so smoothly that the reader could follow it easily and finally reach the places of mathematical interest beyond[1].

This idea of mine is the outcome of many years' teaching in my country. About twenty years ago, when I lectured on a proof of the Lefschetz fixed point theorem in the classroom one day, some students asked me spontaneously:"How did Lefschetz happen to get the idea to formulate his theorem and to give his proof?" Since then, their question came back to my mind whenever I prepared a lecture, and particularly when I read the interesting book [2] in 1973. I began to realize that, in preparing a treatise or a paper as well as a text, it is important for the author to keep in mind the historical development of this subject and the reader's learning process. In 1975 I began to write this book and decided to write it in such a way that it might prove more helpful to my prospective reader. It is the reader who will ultimately pass a reliable judgement on the success or failure of this book.

The material covered is listed in the "Contents", and the long journey from preparation to completion of this book is recorded in the "Epilogue".

Peking University **Kiang Tsai-han**
April 1978

1) This sentence is a translation of the Chinese version which contains some Chinese proverbs: 避免"得意忘形"而使人"望而却步",试图铺平道路以"引人入胜".

Contents

Chapter I

THE GENERAL PROBLEM. A PARTICULAR CASE. A FEW HISTORICAL REMARKS

Introduction

A great number of practical problems have as their mathematical models certain types of mathematical equations, such as systems of linear or algebraic equations, ordinary or partial differential equations, or functional equations. The solution of a system of model equations contributes usually to that of the corresponding practical problem. However, it happens very often in mathematics that the exact solution of a system of equations can be neither determined explicitly nor computed conveniently. Under such circumstances, one has to ask first the following question: Does there exist any solution to the system? Or a deeper question: How many different solutions has the system? After obtaining an affirmative answer to the problem of existence, one proceeds then to look for the exact solutions, or for the practical purpose at hand to make use of certain properties or certain approximations of the exact solutions.

In mathematics, the problem of solving a system of equations can be reduced in general to the problem of determining the fixed points of a self-mapping f of an appropriate space X.[1] Let us denote the set of fixed points of f by $\Phi(f)$ and the number of elements of the set by $\#\Phi(f)$, which is called the *geometric counting* of $\Phi(f)$. This geometric counting is either a nonnegative integer or infinity. On the other hand, when a fixed point of f is located nicely in X (cf. Definitions 4.2, 4.2a), it has the so-called "index", which is a positive or negative integer or null. It was Solomon Lefschetz who first gave

1) By a self-mapping f of a space X we always mean a single-valued continuous mapping f of X to X, i.e., $f\colon X \to X$.

the *algebraic counting* or the *index-sum* of $\Phi(f)$ an algebraic expression $L(f)$, known now as the *Lefschetz number of f* (a positive or negative integer, or null) and proved the following famous theorem (Theorem II 5.1 (iv)).

Theorem *L* (Lefschetz Fixed Point Theorem). *Suppose f is a self-mapping of a connected finite polyhedron X. Then*

(i) *the index-sum of* $\Phi(f)$ *is given by the Lefschetz number* $L(f)$;

(ii) $f' \simeq f \Rightarrow L(f') = L(f)$;

(iii) $L(f) \neq 0 \Rightarrow f$ *has at least one fixed point.*

This is an existence theorem, in which the algebraic counting or the index-sum plays the key role. In view of the conclusion (ii), one is led naturally to introduce the following concept about the geometrical counting——*the least number of fixed points of the mapping class* $\langle f \rangle$ *of* f:

$$\#\Phi(\langle f \rangle) = \text{the largest lower bound of } \{\#\Phi(f') : f' \simeq f\}.$$

The problem of determining $\#\Phi(\langle f \rangle)$ is what we mean by "the general problem" mentioned in our chapter title, and to its discussion our whole book will be devoted.

It was Jakob Nielsen who attacked this problem of determining $\#\Phi(\langle f \rangle)$ with remarkable success. Almost right after Lefschetz's proof of Theorem L, Nielsen introduced the concept of separating $\Phi(f)$ into fixed point classes and the number $N(f)$ of essential (cf. Definition 8.1) fixed point classes (a non-negative integer), now known as the *Nielsen number,* and proved an important case of the following celebrated theorem (Theorem II 6.3):

Theorem *N* (Nielsen Fixed Point Theorem). *Suppose f is a self-mapping of a connected finite polyhedron X. Then*

(i) $f' \simeq f \Rightarrow N(f') = N(f)$;

(ii) $\#\Phi(\langle f \rangle) \geqslant N(f)$, *i.e., f has at least* $N(f)$ *distinct fixed points.*

Our present chapter consists of three parts A, B and C. The first two parts will be devoted to an elementary, detailed, but rigorous treatment of the general problem of determining $\#\Phi(\langle f \rangle)$ for the circle or the 1-sphere S^1, a particular connected finite polyhedron X, including proofs of Theorems L and N for S^1 (Propositions 6.6, 8.4). In Part A, the self-mappings of S^1 considered are the simplest, the so-called integral power mappings f_n, and a formula for $\Phi(f_n)$

is obtained immediately (Proposition 1.2). For the general self-mapping f of S^1 in Part B, there is no more explicit formula for $\Phi(f)$ but an elegant result for $\#\Phi(\langle f \rangle)$ (Proposition 6.8).

As our aim is to present the theory of the fixed point classes for X as a direct generalization of that for S^1, we choose to introduce in Parts A and B definitions of quite a few basic concepts for S^1 and to prove in our own way Theorems L and N for S^1. The definitions and proofs so formulated for S^1 are of course logically simple and intuitively natural, thus paving a way for their generalizations for X in the subsequent chapters. Thus we are able to sketch in Part C the theory of the fixed point classes and its historical development. We hope that the reader may find this rather lengthy chapter helpful in revealing the importance of the natural development from the particular to the general, in rendering the general theory of the fixed point classes easy to comprehend, and in arousing some interests in the applications and development of this theory.

A. *Integral power mappings of the circle*

1. Integral power mappings. The Lefschetz number and fixed points

Let the circle be denoted by

$$S^1 = \{z: |z| = 1\} \quad \text{or} = \{z = e^{2\pi s i}: s \in I\},$$

where z is the complex variable, s the real variable, $i = \sqrt{-1}$ and I the closed interval $[0, 1]$. The simplest and most typical self-mappings of S^1 are the following:

$$f_n: S^1 \to S^1,\ z \mapsto z^n,\ z \in S^1, \text{ or } e^{2\pi s i} \mapsto e^{2\pi n s i},\ s \in I, \tag{1}$$

with integer $n \geqslant 0$. They are called the *integral power mappings.* f_0: $z \mapsto z^0 = 1$ is a constant mapping and f_1 is the identity mapping *id*. When s increases from 0 to 1, the point z starts from the point $z = 1$ on S^1, describes S^1 once in the positive (counterclockwise) sense, and finally returns to the starting point, while the image point z^n has as its starting and terminal points the point $z = 1$, but describes S^1 $|n|$ times in the positive or negative sense according as n is positive or negative.

Because both 0-dimensional and 1-dimensional homology groups of S^1 are infinite cyclic, it is obvious that the corresponding traces in the definition of the Lefschetz number of f_n are 1 and n respectively.

Thus we have

1.1 Proposition. *The Lefschetz number of the integral power mapping f_n of S^1 is $L(f_n) = 1-n$.* □

The following is easy:

1.2 Proposition. *When $n \neq 1$, f_n has $|1-n|$ distinct fixed points given by*

$$\Phi(f_n) = \{z = e^{2\pi ri/(1-n)}: r = 0, 1, \cdots, |1-n|-1\}; \qquad (2)$$

and $\Phi(f_1) = S^1$. □

Remark. $\Phi(f_0) = \Phi(f_2) = \{z=1\}$.

Suppose $n \neq 0, 1, 2$. The $\Phi(f_n)$ as given by (2) consists of $|1-n|$ (>1) equidistant points of division of S^1, including the point $z = 1$ (Figure 1). When r in (2) runs through $0, 1, \cdots, |1-n|-1$, the point $z = e^{2\pi ri/(1-n)}$ runs through these points of division in succession in the positive or negative sense of S^1 according to $n < 0$ or >2.

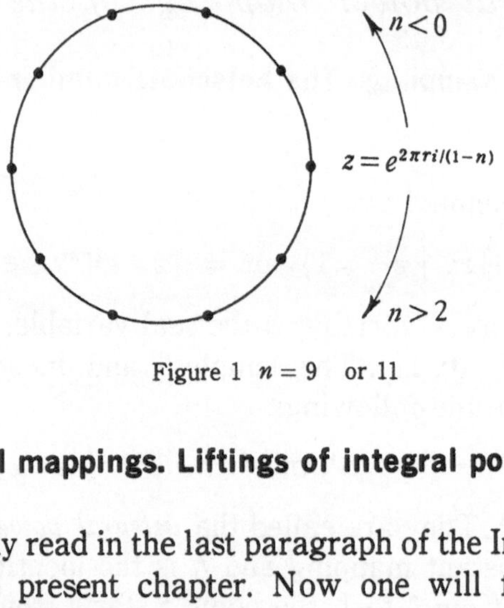

Figure 1 $n = 9$ or 11

2. Exponential mappings. Liftings of integral power mappings

One has already read in the last paragraph of the Introduction a brief outline of the present chapter. Now one will find in §§ 2-3 the motivation of Part B and a prototype of the theory of the fixed point classes in Part B.

The representation of the circle S^1 given by $z = e^{2\pi si}$, $s \in I$, suggests the following mapping

$$p_I: I \to S^1, \ s \mapsto z = e^{2\pi si}, \ s \in I.$$

As the function $e^{2\pi si}$ is single-valued, continuous and periodic with the period 1 on the real line R^1, p_I can be extended to the so-called *exponential mapping*:

$$p: \mathbb{R}^1 \to S^1, \ s \mapsto z = e^{2\pi si}. \tag{1}$$

Obviously p_I is the restriction of p on I, $p(s+k) = p(s)$ for any integral k, and the inverse function is

$$p^{-1}(z) = \frac{1}{2\pi i} \log z = \{s+k: \ z = e^{2\pi si} \text{and } k \text{ is integral}\}, \tag{2}$$

an infinitely multi-valued function. Using the terminology in the theory of covering spaces (cf. Appendix B), we call $\mathbb{R}^1$ the universal covering of S^1, p also the projection and $p^{-1}(z)$ in (2) the fibre at the point $z = e^{2\pi si}$ of S^1. In Figure 2, $\mathbb{R}^1$ is represented as a spiral over S^1, extended indefinitely at both ends, and the fibres at $z=1$ and $z = e^{2\pi si}$ are indicated.

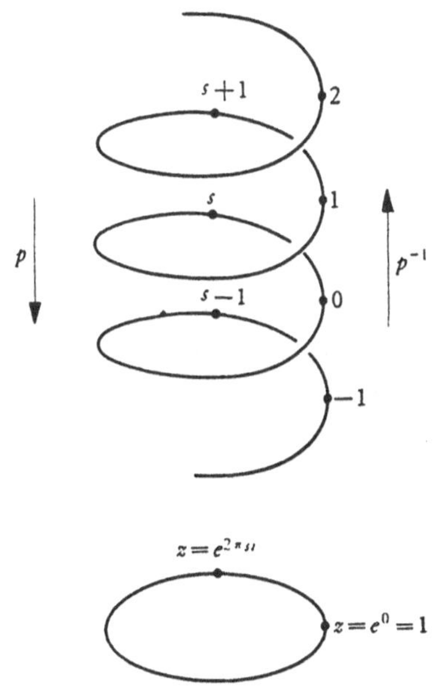

Figure 2

Let us return now to the integral power mapping of S^1 for a given n, $f_n: z \mapsto z^n = e^{2\pi nsi} = e^{2\pi(ns+k)i}$. The infinitely many branches of the multi-valued inverse $p^{-1}(z)$ in (2) correspond to the infinitely many integral values of k. Thus when k is also given, the corresponding branch suggests the following self-mapping:

$$\tilde{f}_{n,k}: \ \mathbb{R}^1 \to \mathbb{R}^1; \ s \mapsto ns + k. \tag{3}$$

The integral value $\tilde{f}_{n,k}(0) = k$ is called the *initial value* of $\tilde{f}_{n,k}$. Independently of the initial value k, there results from (3) the following

$$\tilde{f}_{n,k}(s+1)-\tilde{f}_{n,k}(s)=n, \tag{4}$$

which reveals the basic geometric fact that $f_n(z)$ describes S^1 n times when z goes round S^1 once.

One can prove easily the following

2.1 Proposition. *The mapping $\tilde{f}_{n,k}$ satisfies the functional equation*

$$p\circ\tilde{f}_{n,k}=f_n\circ p. \tag{5}$$

Moreover, if a self-mapping $\tilde{f}$ of R^1 satisfies the functional equation $p\circ\tilde{f}=f_n\circ p$, then $\tilde{f}$ is $\tilde{f}_{n,k}$, where $k=\tilde{f}(0)$. □

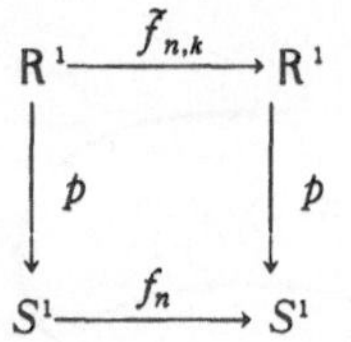

The functional equation (5) is also called the *commutativity relation* of the above diagram. The diagram shows vividly that R^1 covers S^1 and all $\tilde{f}_{n,k}$ cover f_n, or in other words, R^1 is the universal covering space of S^1 defined by p and $\tilde{f}_{n,k}$ are liftings of f_n. This is why one prefers the diagram to (5). Note that f_n has infinitely many liftings given by (3).

3. Fixed points of liftings. Lifting classes and fixed point classes

Being linear functions on R^1, the liftings $\tilde{f}_{n,k}$ are by far simpler than f_n on S^1. It is also much easier to determine $\Phi(\tilde{f}_{n,k})$ than $\Phi(f_n)$. Moreover the following investigation of the relation between them will lead us to the *key concept of the fixed point classes.*

In this section we shall assume that $n\neq 1$ unless otherwise specified. $\tilde{f}_{n,k}$ has only one fixed point, namely,

$$\Phi(\tilde{f}_{n,k})=\{s=k/(1-n)\}\ . \tag{1}$$

For a given pair of integers n $(\neq 1)$ and k, a unique pair of integers q and r, the quotient and the remainder, is determined, such that

$$k=q(1-n)+r,\ 0=r<|1-n|, \tag{2}$$

and thus from Proposition 1.1,

$$p(\Phi(\tilde{f}_{n,k}))=p(k/(1-n))=e^{2\pi ri/(1-n)}\in\Phi(f_n),$$

the p-image of the fixed point of a lifting $\tilde{f}_{n,k}$ of f_n is a fixed point of f_n.

The results on the relation between $\Phi(\tilde{f}_{n,k})$ and $\Phi(f_n)$, so far obtained above, may be summarized by means of the following:

$$\Phi(f_n)=\bigcup_k p(\Phi(\tilde{f}_{n,k})); \tag{3}$$

$$p(\Phi(\tilde{f}_{n,k}))=p(\Phi(\tilde{f}_{n,k'})) \Leftrightarrow k \equiv k' \bmod (1-n); \tag{4}$$

$$p(\Phi(\tilde{f}_{n,k}))\cap p(\Phi(\tilde{f}_{n,k'})) = \emptyset \Leftrightarrow k \not\equiv k' \bmod (1-n). \tag{5}$$

We are thus led to separate the totality of liftings of f_n into disjoint classes, and call

$$[\tilde{f}_{n,k}]= \{\tilde{f}_{n,k'}: k'\equiv k \bmod (1-n)\} \tag{6}$$

the lifting class of f_n, containing the given $\tilde{f}_{n,k}$. Then we have the significant facts that, for $n\neq 1$, f_n has exactly $|1-n|$ lifting classes:

$$[\tilde{f}_{n,r}], \quad r=0,1,\cdots, \ |1-n|-1,$$

and that, by virtue of $p(\Phi(\tilde{f}_{n,r}))=e^{2\pi ri/(1-n)}$, the $|1-n|$ lifting classes of f_n are in one-one correspondence with the $|1-n|$ fixed points of f_n. In Figure 3, the liftings $\tilde{f}_{-2,k}$ of f_{-2} and their separation into 3 lifting classes are shown.

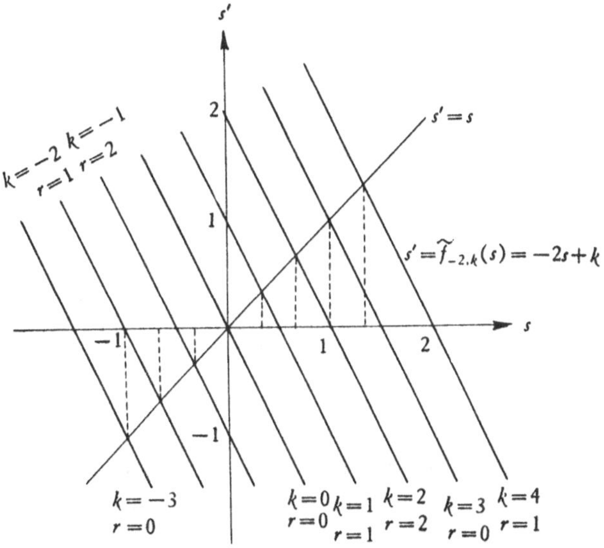

Figure 3 The liftings of f_{-2} and 3 lifting classes

Now let us call $p(\Phi(\tilde{f}_{n,k}))$ a *fixed point class* of $\Phi(f_n)$, when $\tilde{f}_{n,k}$ is any lifting of f_n. From (4), this fixed point class may also be denoted by $p(\Phi([\tilde{f}_{n,k}]))$. The result obtained so far may be restated as the following

3.1 Proposition. *For $n \neq 1$, the set $\Phi(f_n)$ of the $|1-n|$ fixed points of the integral power mapping f_n of S^1 is separated into $|1-n|$ non-empty, and pairwise disjoint, fixed point classes.* □

Consider finally the case $n=1$, namely, $f_n = f_1 = id$. If we interpret $k \equiv k' \bmod (1-n)$ in formulas(3) to (5) as $k=k'$, then (6) shows that

$$[\tilde{f}_{1,k}] = \{\tilde{f}_{1,k}\}$$

consists only of the lifting $\tilde{f}_{1,k}$, and formulas (3) to (5) remain valid for $n=1$. Now let us define for $n=1$ the fixed point classes as before. Then we have

3.2 Proposition. *For $n=1$, i.e. $f_n = f_1 = id$, f_1 has infinitely many distinct lifting clases $[\tilde{f}_{1,k}] = \{\tilde{f}_{1,k}\}$, k being integral, but only one non-empty fixed point class $\Phi(f_1) = p(\Phi(\tilde{f}_{1,0})) = S^1$, and all other fixed point classes are empty.* □

B. *General self-mapping of the circle*

In this Part B, we shall deal with the general self-mapping f of S^1. Although it is no longer possible to obtain $\Phi(f)$ by solving directly the equations $f(z)=z$ and $|z|=1$, but $\# \Phi(\langle f \rangle)$ will be determined in § 6 and Theorem N for S^1 will be proved in § 8. This is accomplished on the basis of the Part A after introducing the *key concept of index* of an isolated fixed point in § 4. The characterization of $\langle f \rangle$ in § 5 and the approximation proposition in § 6 have no counterparts in Part A, while § 7 is analogous to § 3 in defining the fixed point classes via the lifting classes.

4. Index or algebraic counting of an isolated fixed point

It was mentioned that when a fixed point is nicely located in the underlying space, it has an index or algebraic counting. This key notion will be made precise by the following obvious proposition and two definitions (4.2 and 4.2a) for the self-mappings (with $\tilde{f}_{n,k}$ as particular examples) of $\mathbb{R}^1$.

4.1 Proposition. *Suppose s_0 is an isolated fixed point of a self-mapping $\tilde{f}$ of $\mathbf{R}^1$, i.e. there exists a positive number ε such that on the closed interval $[s_0-\varepsilon, s_0+\varepsilon]$ $\tilde{f}$ has the only fixed point s_0. There are exactly the following 4 cases:*

1) $s_0-\varepsilon < \tilde{f}(s_0-\varepsilon)$ *and* $s_0+\varepsilon > \tilde{f}(s_0+\varepsilon)$,
2) $s_0-\varepsilon > \tilde{f}(s_0-\varepsilon)$ *and* $s_0+\varepsilon < \tilde{f}(s_0+\varepsilon)$,
3) $s_0-\varepsilon < \tilde{f}(s_0-\varepsilon)$ *and* $s_0+\varepsilon < \tilde{f}(s_0+\varepsilon)$,
4) $s_0-\varepsilon > \tilde{f}(s_0-\varepsilon)$ *and* $s_0+\varepsilon > \tilde{f}(s_0+\varepsilon)$.   □

Take a rectangular coordinate system in the ss'-plane. Over the interval $[s_0-\varepsilon, s_0+\varepsilon]$ of the s-axis, the curve $s'=\tilde{f}(s)$ and the straight line $s'=s$ intersect only at the point (s_0, s_0). Figure 4 shows us these 4 cases when s increases from $s_0-\varepsilon$ to $s_0+\varepsilon$.

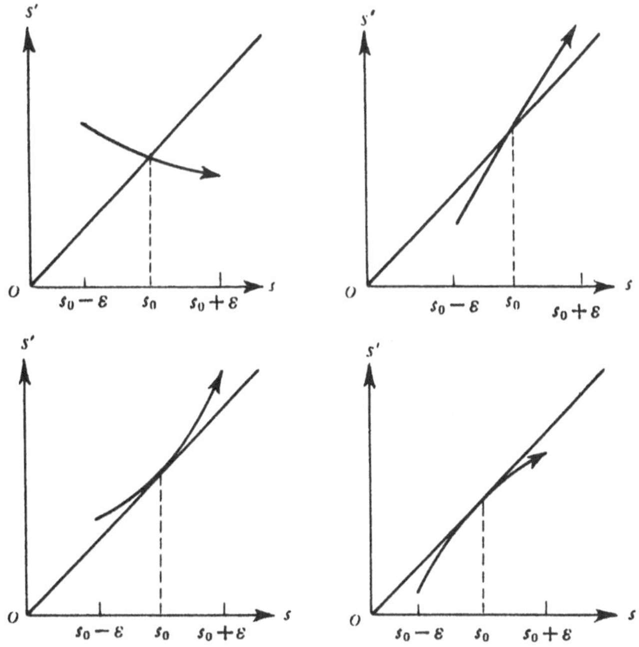

Figure 4

This proposition has another picturesque interpretation, which we adopt as definition of index of an isolated fixed point:

4.2 Definition. Suppose that, under the hypothesis of proposition 4.1, on $\mathbf{R}^1$ the point s moves from $s_0-\varepsilon$ to $s_0+\varepsilon$ and its image point $\tilde{f}(s)$ moves in accordance with the law of $\tilde{f}$. They meet at the fixed point s_0 of $\tilde{f}$. Then (corresponding to the case 1), the case 2), or the cases 3) and 4)), the index of the fixed point s_0 of $\tilde{f}$ is

$$v(\tilde{f},s_0)=\begin{cases} +1, & \text{when the point } s \text{ surpasses the point } \tilde{f}(s) \text{ at } s_0; \\ -1, & \text{when the point } \tilde{f}(s) \text{ surpasses the point } s \text{ at } s_0; \\ 0, & \text{when neither of them surpasses the other at } s_0. \end{cases}$$

Example 4.1. The following self-mapping of R^1

$$\tilde{f}(s)=\begin{cases} s/2, & \text{when } s \leqslant 0; \\ 2s, & \text{when } 0 \leqslant s \end{cases}$$

has the only fixed point $s_0=0$. One can see easily that this fixed point gives rise to the case 4) in Proposition 4.1 and its index is $v(\tilde{f},s_0)=0$.

Example 4.2. The following self-mapping of R^1

$$f(s)=\begin{cases} s+\varepsilon, & \text{for } s \leqslant -\varepsilon \text{ and } \varepsilon \leqslant s; \\ 0, & \text{for} -\varepsilon \leqslant s \leqslant 0; \\ 2s, & \text{for } 0 \leqslant s \leqslant \varepsilon \end{cases}$$

has also only one fixed point $s_0=0$, and $v(\tilde{f},0)=0$.

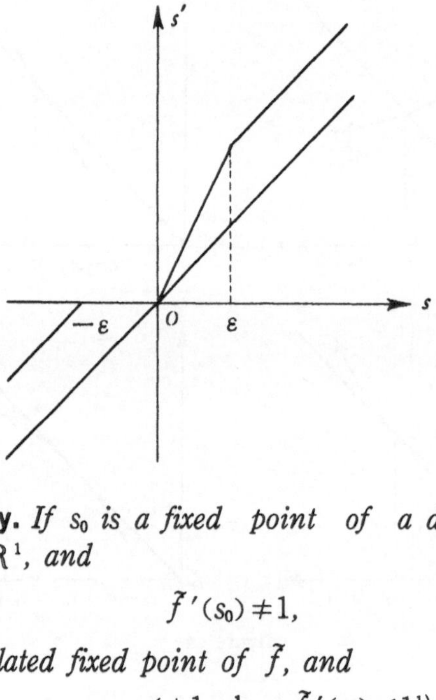

4.3 Corollary. *If s_0 is a fixed point of a differentiable self-mapping $\tilde{f}$ of R^1, and*

$$\tilde{f}'(s_0) \neq 1,$$

then s_0 is an isolated fixed point of $\tilde{f}$, and

$$v(\tilde{f},s_0)=\begin{cases} +1, \text{when } \tilde{f}'(s_0)<1^{1)}; \\ -1, \text{ when } \tilde{f}'(s_0)>1. \end{cases}$$

Proof. Suppose $\tilde{f}'(s_0)<1$. Then for sufficiently small positive ε, we have

1) It is worth while noting that $v(\tilde{f},s_0)=+1$ when $\tilde{f}$ is the constant mapping $\tilde{f}(s)\equiv s_0$.

$$\frac{\tilde{f}(s_0-\varepsilon)-\tilde{f}(s_0)}{-\varepsilon}<1, \quad \frac{\tilde{f}(s_0+\varepsilon)-\tilde{f}(s_0)}{\varepsilon}<1.$$

Since $\tilde{f}(s_0)=s_0$, there follow

$$\tilde{f}(s_0-\varepsilon)>s_0-\varepsilon, \qquad \tilde{f}(s_0+\varepsilon)<s_0+\varepsilon.$$

These show that s_0 is an isolated fixed point of $\tilde{f}$ and that this fixed point gives rise to the case 1) in Proposition 4.1. Thus $v(\tilde{f}, s_0)=+1$.

For the case $\tilde{f}'(s_0)>1$, the proof is similar. □

From this corollary there follows that, for $n\neq 1$, the index of the fixed point of $\tilde{f}_{n,k}$ is

$$v(\tilde{f}_{n,k}, k/(1-n))=\operatorname{sgn}(1-n)^{1)}=\begin{cases}+1, & \text{for } n<1;\\ -1, & \text{for } n>1.\end{cases}$$

It is also the index-sum of the fixed points in $\Phi(\tilde{f}_{n,k})$.

Now let us turn to the definition of an isolated fixed point of a general self-mapping f (with f_n as particular examples) of S^1, which is similar to Difinition 4.2.

4.2a Definition. Suppose z_0 is an isolated fixed point of a self-mapping f of S^1, i. e. there exists on S^1 a closed arc with z_0 as interior point such that on the arc f has the only fixed point z_0. Suppose that the point z describes the arc in the counter-closewise sense and its image $f(x)$ moves on S^1 in accordance with the law of f. They meet at z_0. Then the index of the fixed point z_0 of f is

$$v(f, z_0)=\begin{cases}+1, \text{ when the point } z \text{ surpasses the point } f(z) \text{ at } z_0;\\ -1, \text{ when the point } f(z) \text{ surpasses the point } z \text{ at } z_0;\\ 0, \text{ when neither of them surpasses the other at } z_0.\end{cases}$$

From this definition and the $v(\tilde{f}_{n,k}, k/(1-n))$ obtained above, it follows immediately that, for $n\neq 1$, the index of each of the $|1-n|$ fixed points of the integral power self-mapping f_n is sgn $(1-n)$, and the index-sum of the set $\Phi(f_n)$ is $1-n$.

5. Liftings of a self-mapping. Homotopic self-mapping classes. Fixed points of liftings

As a generaligation of (5) in § 2, we have the following

1) $$\operatorname{sgn} x=\begin{cases}+1, & \text{for} x>0;\\ 0, & \text{for} x=0;\\ -1, & \text{for} x<0.\end{cases}$$

5.1 Definition. Suppose f is a self-mapping of S^1. If a self-mapping $\tilde{f}$ of $\mathbb{R}^1$ is such that the following functional equation

$$p \circ \tilde{f} = f \circ p \tag{1}$$

holds, then $\tilde{f}$ is called a *lifting* of f.

The equation (1) says that the following diagram has commutativity property (see the paragraph after Proposition 2.1)

$$\begin{array}{ccc} s & \overset{\tilde{f}}{\longmapsto} & \tilde{f}(s) \\ \Big\downarrow p & & \Big\downarrow p \\ z = e^{2\pi s i} & \overset{f}{\longmapsto} & f(z) = f(e^{2\pi s i}) = e^{2\pi \tilde{f}(s) i}, \end{array}$$

or in other words, that $2\pi s$ is an argument of the point $z \Rightarrow 2\pi \tilde{f}(s)$ is an argument of $f(z)$. $\tilde{f}(0)$ is also called the *initial value* of $\tilde{f}$. Since $\tilde{f}$ is a single-valued continuous function on $\mathbb{R}^1$, $\tilde{f}(s)$ is obviously a branch of the multivalued function $\frac{1}{2\pi i} \log f(z)$.

5.2 Proposition. *Suppose that f is a self-mapping of S^1. If the point $f(1)$ of S^1 has $2\pi c$ as one of its arguments, then there exists only one lifting $\tilde{f}$ of f with its initial value $\tilde{f}(0) = c$.* □

5.3 Proposition. *Suppose that f is a self-mapping of S^1. If $\tilde{f}$ is a lifting of f, then, for any given integer k, $\tilde{f}(s) + k$ is also a lifting of f. Conversely, if $\tilde{g}$ is another lifting of f, then there is an integer k such that $\tilde{g}(s) = \tilde{f}(s) + k$.*

Proof. The first conclusion is obvious. In order to prove the second conclusion, let $\tilde{f}(0) \equiv c$ and $\tilde{g}(0) = c'$. Since $2\pi c$ and $2\pi c'$ are two arguments of the same point $f(1)$, the difference $c' - c$ is an integer denoted by k. The same consideration shows that, for any s, $\tilde{g}(s) - \tilde{f}(s)$ is an integer. But, on the other hand, $\tilde{g}(s) - \tilde{f}(s)$ is a continuous function, and therefore $\tilde{g}(s) = \tilde{f}(s) + k$. □

5.4 Corollary. *If $\tilde{f}$ is a lifting of f, then there exists an integer n such that for any s*

$$\tilde{f}(s+1) - \tilde{f}(s) = n. \tag{2}$$

Morever, any lifting of f satisfies the above equation (2). □

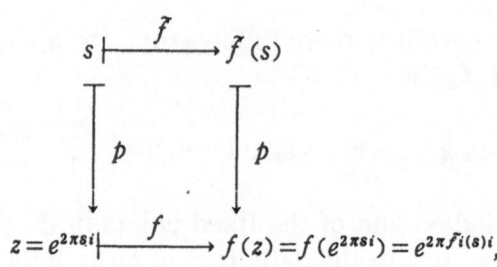

5.5 Definition. When a lifting $\tilde{f}$ of a self-mapping f of S^1 satisfies the equation (2), we then say that the *mapping degree* of f is n.

For example, the mapping degree of the integral power mapping f_n of S^1 is n.

From these propositions, one sees that a self-mapping f of S^1 having a mapping degree n determines a family (with k as parameter) of liftings $\tilde{f}$, which all satisfy the equation (2'), and, conversely, any self-mapping $\tilde{f}$ of $\mathbb{R}^1$ which satisfies the equation (2) determines a self-mapping f of S^1 having mapping degree n. This shows that *self-mapping f of S^1 and any of its liftings, a self-mapping $\tilde{f}$ of $\mathbb{R}^1$, characterize one another.* Since it is complicated to make the graph of f, one often uses as its representative the graph of $\tilde{f}$ on the closed interval $[0, 1]$ of $\mathbb{R}^1$ (Figure 5).

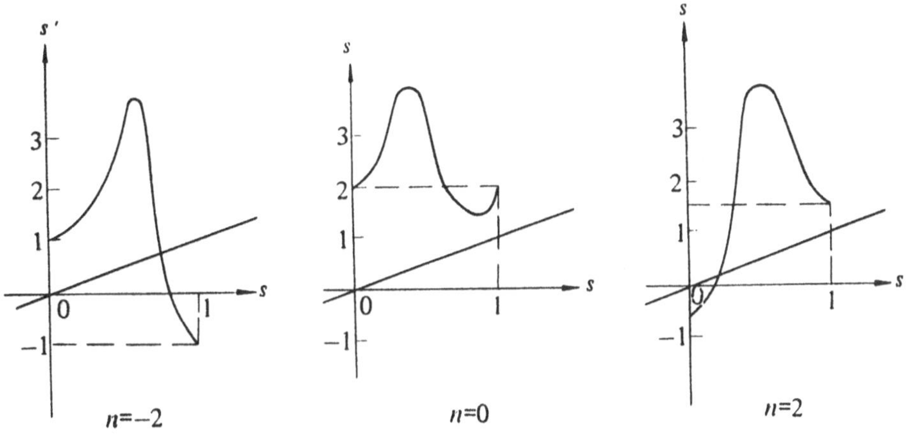

Figure 5 $\tilde{f}$ represents f

The mapping degree of a self-mapping f of S^1 characterizes the homotopy class of f . This is the following proposition, for a proof of which the above remark will be used.

5.6 Proposition (Classification Theorem of Self-Mappings of S^1). *Two self-mappings f and g of S^1 are homotopic $\Leftrightarrow$ they have the same mapping degree.*

Proof. If h_t: $f \simeq g$ is a homotopy between f and g, then the mapping degree of h_t is a continuous function of t; but the value of mapping degree is integral $\Rightarrow$ this function is a constant one, i. e. the mapping degree of $f = h_0$ and $g = h_1$ are the same.

Conversely, suppose the mapping degree of f and g are the same integer n. Then, when $\tilde{f}$ and $\tilde{g}$ are respective liftings of f and g, Definition 5.5 leads to

$$\tilde{f}(s+1)-\tilde{f}(s)=n, \quad \tilde{g}(s+1)-\tilde{g}(s)=n.$$

Let the family of functions be:

$$\tilde{h}_t(s)=(1-t)\tilde{f}(s)+\tilde{t}\,g(s), \quad 0\leqslant t\leqslant 1,$$

which is a homotopy $\tilde{h}_t:f\simeq\tilde{g}$ obviously. Moreover $\tilde{h}_t$ *satisfies the following equation*

$$\tilde{h}_t(s+1)-\tilde{h}_t(s)=n;$$

and hence, for each value of t, $\tilde{h}_t$ determines a self-mapping $h_t:z\mapsto h_t(z)=e^{2\pi\tilde{h}_t(s)i}$ of S^1 with $\tilde{h}_t$ as one of the liftings of h_t. h_t is therefore a homotopy between f and g. □

This proposition implies that if the mapping degree of a self-mapping f of S^1 is n, then $f\simeq f_n$ or $f\in\langle f_n\rangle$, where f_n is the integral power mapping of S^1. It is in this sense that f_n is a typical representative of the general self-mapping f of S. Picturesquely speaking, if a rubber loop were wound around a rigid S^1 in accordance with the law f, then under the action of elasticity it would become a loop wound uniformly around S^1 in accordance with the law f_n.

The following proposition bears the relation between $\Phi(f)$ and $\Phi(\tilde{f})$.

5.7 Proposition. *Let f be a self-mapping of S^1, and $\tilde{f}$ be a lifting of f. Suppose $s_0\in\mathbb{R}^1$, $z_0=p(s_0)\in S^1$. Then*

(i) *s_0 is a fixed point of $\tilde{f}\Rightarrow z_0$ is a fixed point of f;*

(ii) *z_0 is a fixed point of $f\Leftrightarrow s_0$ is a fixed point of a certain lifting*

$$\tilde{g}(s)=\tilde{f}(s)+k \text{ of } f;$$

(iii) *s_0 is a non-isolated fixed point of $\tilde{f}\Rightarrow z_0$ is a non-isolated fixed point of f;*

(iv) *s_0 is an isolated fixed point of $\tilde{f}\Rightarrow z_0$ is an isolated fixed point of f, and*

$$v(f,z_0)=v(\tilde{f},s_0).$$

Proof. (i) and (iii) are obvious consequences according to Definition 5.1.

(ii) Since $2\pi\tilde{f}(s)$ is an argument of $f(z)=e^{2\pi\tilde{f}(s)i}$, and z_0 is a fixed point of $f\Leftrightarrow$ there exists an integer k satisfying $s_0-\tilde{f}(s_0)=k$, i. e. s_0 is a fixed point of $\tilde{g}(s)=\tilde{f}(s)+k$.

(iv) s_0 is an isolated fixed point of $\tilde{f}$ and the function $s-\tilde{f}(s)$ is continuous at $s_0 \Rightarrow$ there exists a small number $\varepsilon>0$ such that $0<|s-\tilde{f}(s)|<1$ when $0<|s-s_0|<\varepsilon$. By virtue of (ii), f has only the fixed point z_0 on the small arc of S^1, with z_0 as its middle point and corresponding to $(s_0-\varepsilon, s_0+\varepsilon)$ of $\mathbb{R}^1$, i.e. z_0 is an isolated fixed point of f. The equality of indices follows immediately from Definitions 4.2 and 4.2a. □

6. Theorem L for the circle

6.1 Proposition. *If a self-mapping of S^1 has only isolated fixed points and is with mapping degree n, then the index-sum of the fixed points of f is $L(f)=1-n$ (Proposition 1.1). Hence, for $n\neq 1$, f has at least one fixed point.*

Proof. Since S^1 is compact, f has only isolated fixed points $\Rightarrow f$ has only a finite number of fixed points. When z describes S^1 once, $f(z)$ describes S^1 n times; therefore (the number of times the point z surpasses the point $f(z)$) minus (the number of times z is surpassed by $f(z)$) must be $1-n$. This is the index-sum by Definition 4.2a. □

When f has no fixed point, i.e. $\Phi(f)=\emptyset$, then we say the index-sum of $\Phi(f)$ is 0. For the case that f has non-isoltaed fixed points, in order to determine the index-sum of $\Phi(f)$, in addition to Definition 4.2a we have to make use of the classical approximation method.

6.2 Definition. Let f be a given self-mapping of S^1 and ε a sufficiently small positive number. A selfmapping g of S^1 is called an ε*-approximation* of f, when there exist a lifting $\tilde{f}$ of f and a lifting $\tilde{g}$ of g such that

$$|\tilde{g}(s)-\tilde{f}(s)|<\varepsilon \quad \text{for any } s\in\mathbb{R}^1.$$

6.3 Proposition. *If g is an ε-approximation of f, then the mapping degree of g is the same as that of f.*

Proof. Suppose on the contrary that the mapping degree of f and g are n and $n+d$, $d\neq 0$ respectively. Then

$$\begin{aligned}\{\tilde{g}(s+1)-\tilde{f}(s+1)\}-\{\tilde{g}(s)-\tilde{f}(s)\}&=\{\tilde{g}(s+1)-\tilde{g}(s)\}\\ -\{\tilde{f}(s+1)-\tilde{f}(s)\}&=n+d-n=d\neq 0.\end{aligned}$$

This gives rise to

$$\lim_{\text{integer } m\to\infty} \{\tilde{g}(s+m) - \tilde{f}(s+m)\} = +\infty,$$

a contradiction to Definition 6.2. □

6.4 Proposition (Approximation Theorem). *For any given self-mapping f of S^1 and any given positive number $\varepsilon > 0$, there exists an ε-approximation g of f, which has only isolated fixed points.*

Proof. Let f be a self-mapping of S^1 with mapping degree n and let $\tilde{f}$ be a lifting of f. It is not difficult to see (cf. Figure 5) that there exists a piece-wise linear single-valued continuous function g on $[0,1]$ (its graph being a broken line consisting of a finite number of straight segments) such that

$$\tilde{g}(0) = \tilde{f}(0),\ \tilde{g}(1) = \tilde{f}(1),\ |\tilde{g}(s) - \tilde{f}(s)| < \varepsilon,$$

and the slope of every segment $\neq 1$. Then extend the definition and obtain $\tilde{g}$ on $\mathbb{R}^1$ by means of the functional equation

$$\tilde{g}(s+1) - \tilde{g}(s) = n.$$

The $\tilde{g}$ so constructed determines a self-mapping $g(z) = g(e^{2\pi si}) = e^{2\pi\tilde{g}(s)i}$ of S^1 with $\tilde{g}$ as a lifting. Evidently g is an ε-approximation of f. The condition that the slope of every segment is not equal to 1, implies that every lifting $\tilde{g}' = \tilde{g} + k$ has only isolated fixed points. By virtue of Proposition 5.7, g has only isolated fixed points. □

6.5 Definition. Let f be any given self-mapping of S^1. For a sufficiently small number $\varepsilon > 0$, let g be an ε-approximation of f and g have only isolated fixed points. Then the index-sum of the fixed points of g is called the *index-sum of the fixed points of f.*

Proposition 6.4 makes sure the existence of ε-approximation g with isolated fixed points, and Propositions 6.1 and 6.3 show that the index-sum of the fixed points of f is independent of the particular ε−approximation g chosen.

When f has only isolated fixed points, then f itself is such an approximation. Moreover, for the identity self-mapping f_1 of S^1, the index-sum is 0, because a small rotation of S^1 around the center has no fixed point and is an ε-approximation of f_1.

This justification of Definition 6.5 gives at once the following generalization of Proposition 6.1:

6.6 Proposition (Theorem L **for the Circle).** *If the mapping degree of a self-mapping f of S^1 is n, then the index-sum of the fixed points of f is $L(f) = 1-n$.* □

This is the goal we aim at in the present section. When $n=1$, i.e. $f \in \langle f_1 \rangle = \langle id \rangle$, the index-sum $L(f)=0$; but the geometric counting $\# \Phi(f)$ of the fixed points of f may be 1, 2, or any natural number, as shown respectively by the following examples.

Example 6.1. Let θ denote the central angle of a point of S^1 in radian measure mod 2π and α denote a given small positive number. Consider the self-mapping of S^1

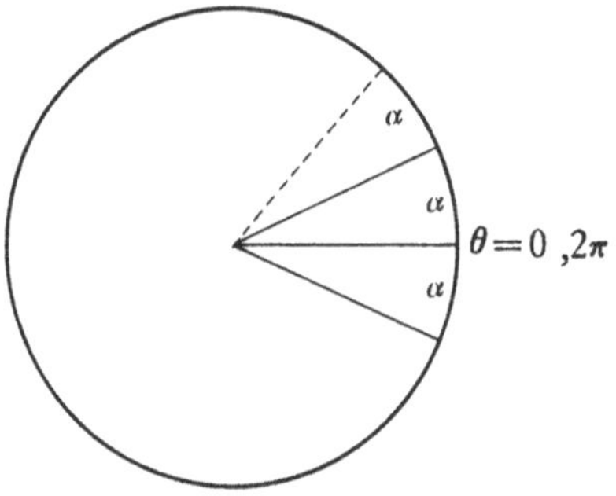

$$f:\ \theta' = \begin{cases} 2\theta, & \text{for } 0 \leqslant \theta \leqslant \alpha; \\ \theta+\alpha, & \text{for } \alpha \leqslant \theta \leqslant 2\pi-\alpha; \\ 2\pi, & \text{for } 2\pi-\alpha \leqslant \theta \leqslant 2\pi. \end{cases}$$

f has $z = 1$ as the only fixed point. From Definition 4.2a, the index of the fixed point is 0. A homotopy between f and id is given by the following

$$f_t:\ \theta' = \begin{cases} 2\theta, & \text{for } 0 \leqslant \theta \leqslant t\alpha; \\ \theta+t\alpha, & \text{for } t\alpha \leqslant \theta \leqslant 2\pi-t\alpha; \\ 2\pi, & \text{for } 2\pi-t\alpha \leqslant \theta \leqslant 2\pi. \end{cases}$$

Example 6.2. Consider the self-mapping of S^1 (with $0<\text{constant}\,\alpha<\frac{\pi}{2}$)

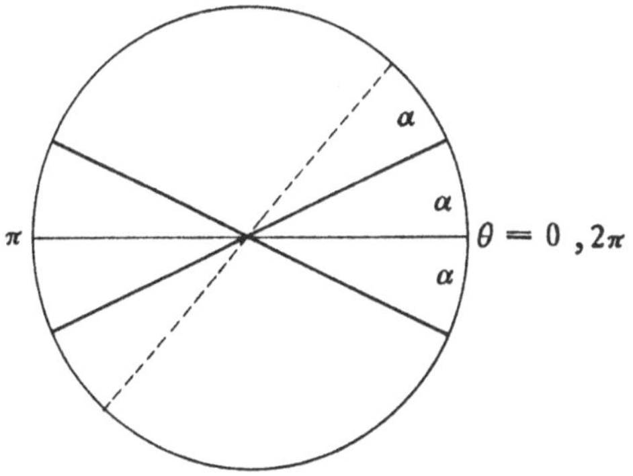

$$f:\theta' = \begin{cases} 2\theta, & \text{for } 0 \leqslant \theta \leqslant \alpha; \\ \theta+\alpha, & \text{for } \alpha \leqslant \theta = \pi-\alpha; \\ \pi, & \text{for } \pi-\alpha \leqslant \theta \leqslant \pi; \\ \pi+2(\theta-\pi) = 2\theta-\pi, & \text{for } \pi \leqslant \theta \leqslant \pi+\alpha; \\ \pi+(\theta-\pi)+\alpha = \theta+\alpha, & \text{for } \pi+\alpha \leqslant \theta \leqslant 2\pi-\alpha; \\ \pi+\pi = 2\pi, & \text{for } 2\pi-\alpha \leqslant \theta \leqslant 2\pi. \end{cases}$$

f has 2 fixed points $z = \pm 1$. Just as in Example 6.1, the index of each of them is 0. On replacing every α in the definition of f by $t\alpha$, we obtain a homotopy $f_t: id \simeq f$.

Example 6.3 (Example in [36 II], p.233, with modification). Consider the self-mapping of S^1 $f: S^1 \to S^1$;

$$e^{2\pi si} \mapsto ie^{2\pi si}e^{\pi iw \sin 2\pi s} = e^{2\pi(s+\frac{w}{2}\sin 2\pi s+\frac{1}{4})i},$$

where w is a given positive integer. Let $f_t: S^1 \to S^1$; $e^{2\pi si} \mapsto e^{2\pi si}e^{2\pi t(\frac{w}{2}\sin 2\pi s+\frac{1}{4})i}$. Then $f_t: id \simeq f_1 = f$, and the mapping degree of f is 1, too. The point $e^{2\pi si}$ is a fixed point of $f \Leftrightarrow \frac{w}{2}\sin 2\pi s + \frac{1}{4} = k$ (an integer); or $z = e^{2\pi si} \in \Phi(f) \Leftrightarrow \frac{w}{2}\sin 2\pi s + \frac{1}{4} = k$ (an integer, $-\frac{w}{2}+\frac{1}{4} \leqslant k \leqslant \frac{w}{2}+\frac{1}{4}$). Now we are to determine the fixed points for given w and k as follows.

Let s_k denote the unique solution of

$$\sin 2\pi s = \frac{2k}{w} - \frac{1}{2w}, \quad -\frac{1}{4} \leqslant s \leqslant \frac{1}{4}, \quad -\frac{w}{2}+\frac{1}{4} \leqslant k \leqslant \frac{w}{2}+\frac{1}{4}.$$

It is easy to see from the accompanying figure that for any integer n

$$\sin 2\pi(n+s_k) = \sin 2\pi(n+\frac{1}{2}-s_k) = \frac{2k}{w} - \frac{1}{2w}.$$

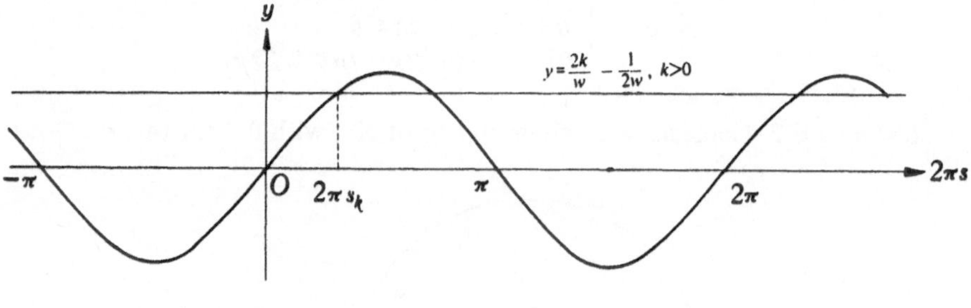

The fixed points are therefore $z_k = e^{2\pi s_k i}$ and $z'_k = e^{2\pi(\frac{1}{2}-s_k)i} = -\frac{1}{z_k}$. They are distinct since $k \neq \pm\frac{w}{2}+\frac{1}{4}$.

The reader may apply Definition 4.2a to determine the indices of the fixed points:

$$v(f, z_k) = -1, \quad v(f, z'_k) = +1.$$

Note that the number of integers k satisfying the relation $-\frac{w}{2}+\frac{1}{4}\leqslant k\leqslant\frac{w}{2}+\frac{1}{4}$ is exactly w, so that f has exactly w pairs of fixed points.

By means of the method developed in the present section, we can obtain not only Proposition 6.6, but also the following

6.7 Proposition. *If a self-mapping f of S^1 has mapping degree n, i.e. f belongs to the homotopy class $\langle f_n\rangle$ of the integral power mapping f_n, then the number*

$$\#\Phi(f)\geqslant|1-n|. \tag{1}$$

Proof. If there are only isolated fixed points, then from Definition 4.2a, the index of any of such fixed points is $+1$, or -1, or 0; and hence (1) follows from the first conclusion of Proposition 6.1. If f has nonisolated fixed point, then $\#\Phi(f)$ is infinite and (1) holds too. □

Remark 1. In this proof we made use of the concept of indices and liftings and Approximation Theorem 6.4. Compare the elementary proof given in [1], p. 533, in which only liftings are involved. We prefer our proof because it leads naturally to its generalization needed in the theory of fixed point classes.

6.8 Proposition. *If a self-mapping f of S^1 has mapping degree n, i.e. $f\in\langle f_n\rangle$, then the least number of fixed points of f*

$$\#\Phi(\langle f\rangle)=|1-n|.$$

Proof. This is an obvious consequence of Propositions 1.2 and 6.7. □

Remark 2. This proposition is not yet Theorem N for S^1, since so far we have not defined $N(f)$. We shall wait until § 8, when Proposition 8.4 which is stronger than Theorem N for S^1 is obtained.

7. Lifting classes. Fixed point classes

For S^1, we already have Theorem L (Proposition 6.6) and a further equality

$$\#\Phi(\langle f\rangle)=|1-n|$$

(Proposition 6.8). Both of them, however, are preliminary to presenting in Part B a prototype of the general theory of fixed point classes, as promised in the last paragraph of the Introduction. The prototype will be completed soon in the following two short sections 7 and 8.

7.1 Notation. Let f be a given self-mapping of S^1, and $\tilde{f}$ a given lifting of f. From Proposition 5.3, any lifting of f must be of the form $\tilde{f}+k$, where k is an integer. As a convention, let $\tilde{f}+k$ be denoted by $\tilde{f}_k$; and hence $\tilde{f}_0$ is the given $\tilde{f}$.

Remark. For convenience, as the given lifting $\tilde{f}$ in Notation 7.1 we take the lifting $\tilde{f}$ with initial value $\tilde{f}(0) \in [0,1]$ (cf. Figure 5).

7.2 Lemma. *Let f be a self-mapping of S^1 with mapping degree n and $\tilde{f}_k$ (Notation 7.1) be the liftings of f. If s is a fixed point of $\tilde{f}_k$ and q is an integer, then $s+q$ is a fixed point of $\tilde{f}_{k'} \Leftrightarrow k'-k=q(1-n)$.*

Proof. From Proposition 5.3 and Corollary 5.4, there follows at once

$$\begin{aligned}\tilde{f}_{k'}(s+q) &= \tilde{f}_k(s+q)+(k'-k) = \tilde{f}_k(s)+nq+(k'-k)\\ &= s+q+\{(k'-k)-q(1-n)\}.\end{aligned}$$ □

The following obvious consequence is needed in the proof of Theorem N for S^1 (Proposition 8.4).

7.2a Corollary. *The hypothesis is the same as that in Lemma 7.2.*

(i) *Case $n \neq 1$. If s is a fixed point of $\tilde{f}_k$ and q a non-null integer, then $s+q$ is not a fixed point of $\tilde{f}_k$. In other words, the projection $p:\Phi(\tilde{f}_k)\to p(\Phi(\tilde{f}_k))$ $(\subseteq \Phi(f))$ is one-to-one.*

(ii) *Case $n=1$. If s is a fixed point of $\tilde{f}_k$, then for any integer q, $s+q$ are all fixed points of $\tilde{f}_k$.* □

The following proposition and definition for a general self-mapping f of S^1 are the immediate generalizations of the corresponding results for the integral power mapping f_n in § 3.

7.3 Proposition. *Let f be a self-mapping of S^1 with mapping degree n, and $\tilde{f}_k$(Notation 7.1) the liftings of f. Then*

(i) $\Phi(f)=\bigcup_k p(\Phi(\tilde{f}_k))$;

(ii) $p(\Phi(\tilde{f}_k))=p(\Phi(\tilde{f}_{k'}))$, *when* $k\equiv k' \bmod (1-n)$;

(iii) $p(\Phi(\tilde{f}_k))\cap p(\Phi(\tilde{f}_{k'}))=\emptyset$, when $k\not\equiv k' \bmod (1-n)$;

where $k\equiv k' \bmod 0$ or $k\not\equiv k' \bmod 0$ for the case $n=1$ are understood as $k=k'$ or $k\neq k'$ respectively.

Proof. (i) is just (ii) of Proposition 5.7, while (ii) and (iii) are obvious consequences of Lemma 7.2. □

7.4 Definition. Let f be a self-mapping of S^1 with mapping degree n and $\tilde{f}_k$ (Notation 7.1) be the liftings of f. For a given integer k, the following subset of liftings

$$\{\tilde{f}_{k'}: k' \equiv k \bmod (1-n)\}$$

is called a *lifting class* of f, denoted by $[\tilde{f}_k]$.

The subset $p(\Phi(\tilde{f}_k))$ of $\Phi(f)$ (independent of the choice of $\tilde{f}_k$ in $[\tilde{f}_k]$ by (ii) of Proposition 7.3) is called a *fixed point class* of f determined by the lifting class $[\tilde{f}_k]$, and is denoted often by $p(\Phi([\tilde{f}_k]))$.

Example 7.1. We continue our Example 6.3. From the definition of f: $S^1 \to S^1$ given there, it is readily seen that the liftings of f are given by

$$\tilde{f}_k(s) = s + \frac{w}{2}\sin 2\pi s + \frac{1}{4} + k.$$

Hence $p(\Phi(\tilde{f}_k)) = \{z_{-k}, z'_{-k}\}$, where z_k is defined as in Example 6.3. On the other hand, since $f \simeq id$, f has the mapping degree 1. Hence by Definition 7.4, each lifting $\tilde{f}_k$ is a lifting class by itself. Thus f has exactly w nonempty fixed point classes. Note that w is any given position integer.

Combination of Propositions 7.2a and 7.3 gives readily the following generalization of Propositions 3.1 and 3.2.

7.5 Proposition. *Let f be a self-mapping of S^1 with mapping degree n. The set of all liftings of f is separated into pairwise disjoint liftings classes, and $\Phi(f)$ is separated into pairwise disjoint fixed point classes. For $n \neq 1$, there are $|1-n|$ nonempty fixed point classes; while for $n = 1$, there are infinitely many lifting classes, but the number of nonempty fixed point classes can attain any non-negative integral value (cf. Example 7.1).* □

Note: A lifting class can never be empty, while a fixed point class may be empty. Two different lifting classes may determine the same empty fixed point class.

8. Index of a fixed point class. The Nielsen number. Theorem N for the circle

For the time being, consider self-mapping f of S^1 with only isolated fixed points.

8.1 Definition. Let f be a self-mapping of S^1 with isolated fixed points only. The index-sum of all fixed points of a fixed point class is called the *index of the fixed point class*. A fixed point class is called an *essential class* if its index $\neq 0$. The number of the essential classes is called the *Nielsen number* of f, denoted by $N(f)$.

8.2 Lemma. *If a self-mapping f of S^1 has mapping degree n, i.e. $f \in \langle f_n \rangle$, and has only isolated fixed points, then the index of every fixed point class is* $\operatorname{sgn}(1-n)$. *Therefore*

$$N(f) = |1-n|.$$

Proof. Since S^1 is compact and f has only isolated fixed points, f has only a finite number of fixed points.

Consider first the case $n<1$. From Proposition 5.7 and (i) of Corollary 7.2a, we find respectively that any lifting $\tilde{f}_k$ has also only isolated fixed points, and that the number of fixed points is finite and $\#\Phi(\tilde{f}_k) = \# p(\Phi(\tilde{f}_k))$. From (iv) of Proposition 5.7 and Definition 8.1, the index-sum of $\Phi(\tilde{f}_k)$ is the index of the fixed point class $p(\Phi(\tilde{f}_k))$ determined by $\tilde{f}_k$.

Thus our problem is to determine the index-sum of $\Phi(\tilde{f}_k)$. Form the function

$$\varphi(s) = s - \tilde{f}_k(s). \tag{1}$$

Since the mapping degree of f is n, there follow from Corollary 5.4 and Definition 5.5

$$\tilde{f}_k(s+1) - \tilde{f}_k(s) = n, \tag{2}$$

$$\varphi(s+1) - \varphi(s) = 1-n > 0; \tag{3}$$

and hence

$$\lim_{s \to \pm\infty} \varphi(s) = \pm\infty,$$

and (1) gives respectively $s \gtrless \tilde{f}_k(s)$. This tells us that

index-sum of $\Phi(\tilde{f}_k)$
= (the number of times the point s surpasses the image point $\tilde{f}_k(s)$) − (the number of times s is surpassed by $\tilde{f}_k(s)$)
= 1,

and therefore the index of every fixed point class of f is 1.

The case $n>1$ can be treated similarly, and the conclusion now is that the index of every fixed point class of f is -1.

It remains to consider the case $n=1$. Now (2) and (3) are replaced respectively by

$$\tilde{f}_k(s+1) - \tilde{f}_k(s) = 1, \tag{4}$$

$$\varphi(s+1) - \varphi(s) = 0. \tag{5}$$

From (5), the function $\varphi(s)$ is periodic with the period $=1$. Since f

has only isolated fixed point, by virtue of (iii) of Proposition 5.7, so has $\tilde{f}_k$; then there exists a point s_0 which is not a fixed point of $\tilde{f}_k$. From (ii) of Corollary 7.2a, s_0+1 is also not a fixed point of $\tilde{f}_k$. This result combining with (5) gives

$$\varphi(s_0+1)=\varphi(s_0)\neq 0. \tag{6}$$

Denote by $\Phi'(\tilde{f}_k)$ the subset of $\Phi(\tilde{f}_k)$ on the period interval $[s_0,s_0+1]$ of φ. From (ii) of Corollary 7.2a again, the projection p: $\Phi'(\tilde{f}_k)\to p(\Phi(\tilde{f}_k))$ is one- to-one, and hence by virtue of (6)

$$\begin{aligned}&\text{index of } p(\Phi(\tilde{f}_k)) \;=\; \text{index-sum of } \Phi'(\tilde{f}_k)\\&=(\text{the number of times } s \text{ surpasses } f_k(s))\\&\quad-(\text{the number of times } s \text{ is surpassed by } \tilde{f}_k(s))\\&=0.\end{aligned}$$

□

We shall define $N(f)$ for the case that f has also non-isolated fixed points, by using approximation method just as in Definition 6.5.

8.3 Definition. Let f be a self-mapping of S^1. By virtue of Proposition 6.4, for any sufficiently small $\varepsilon>0$, there exists ε-approximation g of f such that g has only isolated fixed points. The Nielsen number $N(g)$ of g is called the *Nielsen number* of f and is denoted by $N(f)$.

By virtue of Proposition 6.3 and Lemma 8.2, $N(f)$ so defined is independent of the g chosen. In case f has originally the isolated fixed points only, f itself may be taken as g; then this Definition is in agreement with Definition 8.1. Combining Lemma 8.2 and Definition 8.3 and recalling Proposition 6.8, we arrive at the final goal:

8.4 Proposition (Stronger than Theorem *N* for the Circle). *If a self-mapping f of S^1 has mapping degree n, then*

$$N(f)=|1-n|;$$

and moreover

$$\#\Phi(\langle f\rangle)=N(f).$$

□

Remark. The second conclusion of our proposition is an equation, not the same as the conclusion of Theorem N in the Introduction.

C. *A sketch of the theory of fixed point classes. A few historical remarks*

9. From particular case to theory of fixed point classes

Let f be a self-mapping of a finite connected polyhedron X and $\#\Phi(\langle f\rangle)$ the least number of the fixed points of the homotopy class $\langle f\rangle$ of f. In the Introduction of the present chapter we mentioned that our book would aim at the general problem of determining $\#\Phi(\langle f\rangle)$ and our present Part C in particular at a sketch of the general theory of the fixed point classes.

Let us now recall what we have done in parts A and B. In brief outline, for the particular case $X = S^1$ we have introduced before the proof of Theorem N the following important concepts: 1) homotopy class or mapping class $\langle f\rangle$, 2) the Lefschetz number $L(f)$, 3) lifting $\tilde{f}$, 4) lifting class $[\tilde{f}]$, 5) index $v(f,x_0)$ of an isolated fixed point x_0 of f, 6) fixed point class $p(\Phi(\tilde{f}))$ and its index (denoted hereafter by F_i and v_i respectively) and 7) the Nielsen number $N(f)$.

As we have promised that the presentation of the particular case S^1 in Parts A and B is such that the theory of the fixed point classes for a general finite connected polyhedron X will be its direct generalization, one will expect that these seven concepts can be formulated for the general case X also. In fact, the first two for X can be found in elementary texts on topology, while the last five for X will be given in next chapter, on the basis of the more effective theories of covering spaces (Appendix B) and indices of fixed points (Theorem II 5.1). More precisely, making use of X and its universal covering $\tilde{X}$ in the next chapter in the same way just as we did previously with S^1 and $\mathbb{R}^1$, we shall prove Nielsen's classical results: $N(f)$ is finite; $N(f)$ and the index v_i of F_i are homotopy invariants; and finally the Nielsen fixed point theorem is obtained.

10. A few historical remarks

Let $f: X \to X$ be again a self-mapping of finite connected polyhedron X. In the fixed point theory there have been two apparently different problems, namely the existence problem and the problem of the least number of fixed points. Nielsen's concept of fixed point classes of about half a century records ([21] , [32I]) marked the beginning of studying the latter problem. These two problems are in fact closely related. We have seen that the discovery of $L(f)$ preceded to that of $N(f)$ by only a few years, and we shall see that via $N(f)$ there results a converse theorem of the Lefschetz fixed point theorem (Theorem 5.4): If X satisfies certain conditions, then $L(f)=0 \Rightarrow$ there exists in $\langle f \rangle$ a self-mapping g of X free from fixed point.

(1) *The problem of existence or non-existence of fixed points*

The problem of solving an equation is not only equivalent in general to the problem of determining the fixed points of a self-mapping, as mentioned in the Introduction, but in fact the fixed point theory has its origin in the former. In Gauss' proof (1799) of the fundamental theorem of algebra, there appeared already the topological concepts of homotopy and multiplicity of a root. As another example we take a theorem proved by H. Poincaré in his famous *Mémoire sur les courbes définies par les équations différentialles* (1881—1886). He introduced the concept of index of an isolated singularity of a vector field on a closed orientable surface M, or in other words, the index of an isolated fixed point of a self-mapping homotopic to the identity. The theorem of Poincaré we have in mind states: the index-sum of the singularities of a vector field with only isolated singularities on M of genus p is $\chi(M)=2-2p$, the Euler-Poincaré charactistic of M. This is evidently a particular consequence of Theorem L. In the beginning of this century, the Dutch mathematician Brouwer introduced the concept of mapping degree of a mapping from one manifold to another of the same dimension, generalized the concept of index from surfaces to manifolds of higher dimensions, and proved (1910) the well-known Brouwer fixed point theorem. His another important result states: if two homotopic self-mappings of a closed n-dimensional manifold have only isolated fixed points, then their index-sums of fixed points equal. This leads naturally to the concept of homotopic invariance of the index-sum.

As a climax of development along this line, Theorem L was discovered by the American mathematician S. Lefschetz. It is an existence theorem of far-reaching significance. His first version [28] is limited to the orientable closed manifold X, but H. Hopf [24] generalized it from manifold to homogeneous[1], finite and connected complex and proved Theorem L in a much easier way.

---

1) An *n-dimensional complex is homogeneous,* when each of its simplexes is a face of at least one of its n-simplexes.

Theorem L has various generalizations. They arise through altering the hypothesis on the space X and/or that on the self-mapping f, as demanded by the mathematical problems under consideration. For example, in the theory of differential equations and in functional analysis X is usually supposed to be a topological space of more general types, while in the theory of numerical analysis and in practical computation of fixed points f is usually supposed to be contractive or compact. They are all beyond the scope of our book, and readers who are interested in them may be referred to [2*], [4], [6], [7], [13], [14].

For the existence of at least one fixed point of a self-mapping $f: X \to X$, the condition $L(f) \neq 0$ is sufficient but not necessary. An example to this effect is the following: X is the figure 8, and f keeps the crossing point fixed and maps the upper and lower loops just as the integral power mappings f_{-1} and f_2 respectively. Now $L(f)=0$, but f has exactly two fixed points and every self-mapping in $\langle f \rangle$ has at least two fixed points (cf. Proposition 8.4); see [1] p. 534, or [2] pp. 28—30, 72—73, 111—112. This leads naturally to the important question: Under what conditions will a converse theorem of Theorem L hold? i.e., under what conditions on X and f, will there be a self-mapping in $\langle f \rangle$ free from fixed point? We shall see that such a converse theorem of Theorem L (Theorem IV5.4 or [2], p. 143) is obtained through the intervention of the concept of the Nielsen number and the least number of fixed points of f.

(2) *The problem of the least number of fixed points*

It is only four years after Lefschetz's [28] that there appeared Danish mathematician J. Nielsen's [32I], which marked the beginning of the theory of the fixed point classes. He dealt only with the particular case of closed orientable surfaces and the self-mappings homotopic to homeomorphisms, and made use of the fundamental groups of the surfaces instead of the homology groups. As Hopf proceeded forward from self-mapping of closed orientable manifold to that of homogeneous finite connected polyhedron, the German mathematician F. Wecken ([36]) proceeded further to self-mapping of general finite connected polyhedron. He proved for the general case those fundamental theorems in the theory of the fixed point classes (see Chapter II), first formulated by Nielsen for the case of orientable closed surface. It is especially worth mentioning that Wecken studied in connection with a self-mapping a series of subgroups of the fundamental group of the polyhedron, which paved the way for further development of this theory.

In the earlier treatise [1] and text [11] the theory of fixed point classes was mentioned. After the publication of Wecken's fundamental work, the French mathematician J. Leray ([29]) pointed out the application of this theory to the theory of partial differential equations, and the English mathematician M. H. A. Newman ([31]) emphasized the role the fundamental group played in Nielsen's work and suggested that further use of the fundamental group and homotopy groups of higher dimensions should lead to substantial development of the fixed point theory. Both these helped to popularize the theory of fixed point classes.

However, up to 1962, we knew the Nielsen number of a self-mapping f of a connected finite polyhedron X only in very simple cases such as 1) $N(f) \leqslant 1$ for simply-connected X, 2) $N(f) \leqslant 1$ for $f \in \langle id \rangle$ (see Example II6.1) and 3) when X is a torus or a lens space. It is our young mathematician B. J. Jiang (sometimes romanized as P. C. Chiang) who took a first step forward in [25]. He started with the concept of liftings of self-mappings (Definition II 1.1) and used with success a subgroup of $\pi_1(X)$, now called *Jiang group* and denoted by $J(f)$. He proved among other results the following: When $J(f) = \pi_1(X)$, then the indices of all fixed point classes equal, and $N(f)$ can be computed by means of $L(f)$. This is an extreme case, and the general problem of estimation of $N(f)$ is far from completely solved (cf. [25 II]). However, his result is applicable to the important case of self-mappings of Lie groups. Our another young mathematician G. H. Shi ([35]) improved and simplified Wecken's condition. His condition is sufficient for $\# \Phi(\langle f \rangle) = N(f)$. It happens that under both Jiang's and Shi's conditions in the two theorems just mentioned, a converse theorem of Theorem L (Theorem III 5.4) is established. For the identity mapping class of a general connected finite polyhedron Shi ([35 a]) obtained a combinatorial method to compute $\# \Phi(\langle id \rangle)$ (see Theorem IV 3.5). Chapters III and IV are devoted respectively to these results of theirs. Among the papers which are closely related to their work and have been published in the interval (1964—1979), two will be presented in brief outline in our last chapter, Chapter V (cf. the bibliography in [26b]).

Chapter II

THE NIELSEN NUMBER

In this chapter we begin to consider the general case of a connected finite polyhedron X and an arbitrary self-mapping f of X.

In the previous chapter, we studied the particular case of the circle S^1 and an arbitrary self-mapping f of S^1. For this particular case, we separated the fixed point set $\Phi(f)$ of f into fixed point classes by investigating the fixed point sets $\Phi(\tilde{f})$ of all the liftings $\tilde{f}$ of f, and introduced two concepts: the index of a fixed point class and the Nielsen number $N(f)$. Because the theory of universal covering spaces (Appendix B) and the theory of fixed point index (Theorem II 5.1) are available, in the present chapter we can and shall proceed to treat rigorously the general case in a quite parallel manner. The former theory enables us to introduce in §§ 1—2 the concept of fixed point classes of the self-mapping f of the general X, and to discuss in §§ 3—4 the correspondence between the fixed point classes of f and those of g, induced by a homotopy $f \simeq g$. The latter theory enables us to formulate in §§ 5—6 the definitions of the index of a fixed point class and of the Nielsen number, to prove their invariance under homotopy, and to obtain finally the Nielsen fixed point Theorem 6.3. In § 7, the last section, we shall present our results on the commutativity (Corollary 7.4) and the homotopy type invariance (Theorem 7.8) of the Nielsen number.

1. Lifting class and fixed point class

A connected finite polyhedron X has universal covering space. Note that we mean by $(\tilde{X},p)$ or $\tilde{X}$ *always* the universal covering space of X unless otherwise specified. For all preliminaries about the theory of covering spaces we need, we refer readers to Appendix B. Because the concept of liftings of f plays a special role in our presentation of the theory of fixed point classes, for convenience we repeat here its definition with added remarks and collect the useful results on liftings in a lemma (Lemma 1.2).

1.1 Definition. Let f be a self-mapping of X. A self-mapping $\tilde{f}$ of $\tilde{X}$ is called a *lifting of* f, if for any $\tilde{x} \in \tilde{X}$, the relation

$$p \circ \tilde{f}(\tilde{x}) = f \circ p(\tilde{x})$$

holds, or in other words, if $\tilde{f}$ carries each fibre $p^{-1}(x)$ into the fibre $p^{-1}(f(x))$ (cf. Definitions I 5.1 and B 2.3).

$$\begin{array}{ccc} \tilde{X} & \xrightarrow{\tilde{f}} & \tilde{X} \\ p \downarrow & & \downarrow p \\ X & \xrightarrow{f} & X \end{array}$$

It is then obvious that the covering motions of $\tilde{X}$ *are the liftings of the identity mapping of* X (*cf. Definition* B 3.3 *and Remark* 1 *there*).

1.2 Lemma. *Let* $(\tilde{X},p)$ *be the universal covering space of a connected finite polyhedron* X.

(i) *For an arbitrary point* $x_0 \in X$ *and any two points* $\tilde{x}_0, \tilde{x}'_0 \in p^{-1}(x_0)$, *there is a unique covering motion* $\gamma: \tilde{X} \to \tilde{X}$ *such that* $\gamma(\tilde{x}_0) = \tilde{x}'_0$ (*Corollary* B 5.2). *The totality of all covering motions of* $\tilde{X}$ *form a group, called the group of covering motions and denoted by* $\mathscr{D}(\tilde{X}, p)$ *or simply by* $\mathscr{D}$ (*cf. Theorem* B 3.11 *and* B 3.12).

$$\begin{array}{ccc} \tilde{X}, \tilde{x}_0 & \overset{\gamma}{\dashrightarrow} & \tilde{X}, \tilde{x}'_0 \\ p \downarrow & & \downarrow p \\ X, x_0 & \xrightarrow{id} & X, x_0 \end{array}$$

(ii) *Let* $f: X \to X$ *be a self-mapping of* X. *For any point* $x_0 \in X$, *denote* $x_1 = f(x_0)$. *Then for an arbitrary point* $\tilde{x}_0 \in p^{-1}(x_0)$ *and any*

$$\begin{array}{ccc} \tilde{X}, \tilde{x}_0 & \overset{\tilde{f}}{\dashrightarrow} & \tilde{X}, \tilde{x}_1 \\ p \downarrow & & \downarrow p \\ X, x_0 & \xrightarrow{f} & X, x_1 \end{array}$$

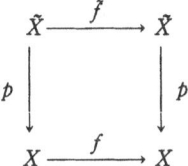

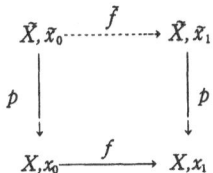

point $\tilde{x}_1 \in p^{-1}(x_1)$, *there exists just a unique lifting* $\tilde{f}: \tilde{X} \to \tilde{X}$ *such that* $\tilde{f}(\tilde{x}_0) = \tilde{x}_1$ (*cf. Corollary* B 5.3).

(iii) *If* $\tilde{f}$ *is a lifting of* f *and* $\alpha, \beta \in \mathscr{D}$, *then* $\beta \circ \tilde{f} \circ \alpha$ *is also a lifting of* f.

(iv) *If both* $\tilde{f}$ *and* $\tilde{f}'$ *are liftings of* f, *then there is a unique* $\gamma \in \mathscr{D}$ *such that* $\tilde{f}' = \gamma \circ \tilde{f}$. □

What we are intersted in now is the problem of fixed points. By way of studying the relation between the fixed points of f and those of the liftings of f, we are led step by step to the concept of lifting classes.

From Definition 1. 1, we have obviously

(a) Let $\tilde{x} \in p^{-1}(x)$. Then $f(x) = x$ or $\neq x \Rightarrow$ for every lifting $\tilde{f}$ of f, $\tilde{f}(\tilde{x}) \in$ or $\notin p^{-1}(x)$ respectively. □

This means, in other words, that according as the point x is, or is not, a fixed point of f, the restriction of lifting $\tilde{f}$ on $p^{-1}(x)$ is or is not a self-mapping of $p^{-1}(x)$.

Moreover, when $f(x) = x$ and $\tilde{f}(\tilde{x}) \in p^{-1}(x)$, there are two more alternatives:

$$\tilde{f}(\tilde{x}) = \tilde{x}, \text{ and } \tilde{f}(\tilde{x}) \neq \tilde{x}.$$

When the latter is the case, by virtue of (ii) of Lemma 1. 2, there exists a unique lifting $\tilde{f}'$ such that $\tilde{f}'(\tilde{x}) = \tilde{x}$. Hence

(b) $f(x) = x \Leftrightarrow$ for any given $\tilde{x} \in p^{-1}(x)$, there is a unique lifting of f, which leaves $\tilde{x}$ fixed. □

This result reduces the problem of determining the fixed point set $\Phi(f)$ to the problem of determining first the fixed point sets $\Phi(\tilde{f})$ of all liftings $\tilde{f}$ of f and then their p-images. Moreover, this result gives rise to the following question: If $\tilde{f}$ has a fixed point $\tilde{x} \in p^{-1}(x)$, which other liftings also have fixed points on $p^{-1}(x)$? By virtue of (i), (ii) and (iii) of Lemma 1.1, the answer is the following:

(c) Let $\tilde{f}$ be a lifting of f with a fixed point $\tilde{x} \in p^{-1}(x)$, and suppose that $\gamma \in \mathscr{D}$ is any covering motion of $\tilde{X}$. The lifting $\gamma \circ \tilde{f} \circ \gamma^{-1}$ of f, then, has a fixed point at $\gamma(\tilde{x}) \in p^{-1}(x)$. Conversely, if a lifting $\tilde{f}'$ of f has a fixed point at $\gamma(\tilde{x}) \in p^{-1}(x)$, then $\tilde{f}' = \gamma \circ \tilde{f} \circ \gamma^{-1}$. □

(c) leads to the following two definitions and two theorems.

1.3 Definition. Let $\tilde{f}$ and $\tilde{f}'$ be two liftings of the same self-mapping f of X. If there is a covering motion $\gamma \in \mathscr{D}$ such that

$$\tilde{f}' = \gamma \circ \tilde{f} \circ \gamma^{-1},$$

we say $\tilde{f}$ and $\tilde{f}'$ are related and denoted their relation by

$$\tilde{f} R \tilde{f}'.$$

This relation is an equivalence relation, and hence all the liftings of f are separated into pairwise disjoint equivalent sets, called the *lifting classes*. Denote the lifting class containing $\tilde{f}$ by

$$[\tilde{f}] = \{\gamma \circ \tilde{f} \circ \gamma^{-1};\ \gamma \in \mathscr{D}\}.$$

1.4 Theorem. *Let $\tilde{f}, \tilde{f}', \cdots$ be liftings of the self-mapping f of X. Then among the fixed point sets $\Phi(f), \Phi(\tilde{f}), \Phi(\tilde{f}'), \cdots$ the following hold:*

$$\Phi(f) = \bigcup_{\tilde{f}} p(\Phi(\tilde{f})), \text{ union over all liftings } \tilde{f} \text{ of } f, \tag{1}$$

$$[\tilde{f}] = [\tilde{f}'] \Rightarrow p(\Phi(\tilde{f})) = p(\Phi(\tilde{f}')), \tag{2}$$

$$[\tilde{f}] \neq [\tilde{f}] \Rightarrow p(\Phi(\tilde{f})) \cap p(\Phi(\tilde{f}')) = \emptyset. \tag{3}$$

Proof. (1) follows from (b). The first conclusion of (c) $\Rightarrow \Phi(\gamma \circ \tilde{f} \circ \gamma^{-1}) = \gamma(\Phi(\tilde{f})) \Rightarrow$ (2), and similarly the second conclusion of $(c) \Rightarrow$ (3). □

This theorem is a generalization of Proposition I 7.3 and leads naturally to Nielsen's definition of the fixed point class ([32I] , p. 289).

1.5 Definition. The subset $p(\Phi(\tilde{f}))$ of the fixed point set $\Phi(f)$ is called a *fixed point class* of f , determined by the lifting $\tilde{f}$ of f, or (by virtue of (2) of Theorem 1.4) by the lifting class $[\tilde{f}]$ of f.

With this definition, we may restate Theorem 1.4 as follows

1.6 Theorem. *The fixed point set $\Phi(f)$ of f is separated into pairwise disjoint fixed point classes $p(\Phi(\tilde{f}))$, $\tilde{f}$ running over all the liftings of f.* □

It is extremely important to note that a lifting class is never empty while a fixed point class may be empty (cf. Proposition I 7.3). In general, one or several fixed point classes determined by different lifting classes may be empty; a lifting class determines a unique fixed point class, while an empty fixed point class may be determined by different lifting classes; when one fixed point class is the whole space X, then all the other fixed point classes must be empty. All these of course contradict neither (3) of Theorem 1.4 nor Theorem 1.6. We are thus led to the following notation and definition.

1.7 Notation. The pair $(p(\Phi(\tilde{f})), [\tilde{f}])$ is called a *labelled fixed point class.* By this we mean the fixed point class $p(\Phi(\tilde{f}))$ with the *label* $[\tilde{f}]$. Henceforth, by a fixed point class we always mean a labelled fixed point class, and we often write $p(\Phi(\tilde{f}))$ in stead of $(p(\Phi(\tilde{f})), [\tilde{f}])$.

1.8 Definition. The number of lifting classes $[\tilde{f}]$ of f, or the number of labelled fixed point classes $(p(\Phi(\tilde{f})), [\tilde{f}])$ of f, is called the *Reidemeister number* of f, denoted by $R(f)$.

The Reidemeister number $R(f)$ may be either a natural number or infinity.

Remark 1. If X is simply-connected, then the universal covering space $\tilde{X}$ of X is X itself, or (X, id). Then any self-mapping f of X has only one lifting $\tilde{f} \equiv f$; and hence $R(f)=1$.

Remark 2. By $(\tilde{X},p)$ we have meant so far in the present chapter the universal covering space of X. There arises naturally the following question: If $(\tilde{X}, p)$ is a covering space of a certain type, but not necessarily the universal covering space, can we formulate in some similar manner the definition of lifting classes and then that of fixed point classes? Let us recall that the properties of the universal covering space on which the concepts of covering motions and liftings have been based are (i) and (ii) of Lemma 1.2. If a covering space $(\tilde{X},p)$ is not necessarily universal, but has the property (i) and the self-mapping f of X has the property (ii), the discussion in the present section can be carried through. From the theory of covering spaces, a regular covering space $(\tilde{X},p)$ of X (i.e. a covering space of X determined by a normal subgroup $H=p_\pi(\pi_1(\tilde{X}, \tilde{x}_0))$ of $\pi_1(X, x_0)$) has the property (i) (see Theorem B 3.11 and B 3.12), and a self-mapping f of X with $f_\pi(H)\subseteq H$ for a normal subgroup H of $\pi_1(X, x_0)$ has the property (ii) (see Theorem B2.4). We shall see in V § 1 that for such a covering space $(\tilde{X},p)$ and such a self-mapping f, certain fixed point classes and the Nielsen number of a more general type are obtained, and are called the fixed point H-classes and the Nielsen number $N(f; H)$.

2. Nonempty fixed point class. Equivalent definition. Finiteness

Leaving the empty fixed point class aside for the moment, we turn to the nonempty fixed point classes, and prove the following very important theorem.

2.1 Theorem. *Two fixed points x_0 and x_1 of a self-mapping f of a connected finite polyhedron X belong to the same fixed point class* *of f $\Leftrightarrow$ there exists in X a path c from x_0 to x_1 such that*

$$f\circ c \simeq c.$$

Proof. Suppose that $\tilde{f}$ is a lifting of f and $x_0 \in p(\Phi(\tilde{f}))$. Then there is $\tilde{x}_0 \in p^{-1}(x_0)$ such that $\tilde{f}\ (\tilde{x}_0) = \tilde{x}_0$.

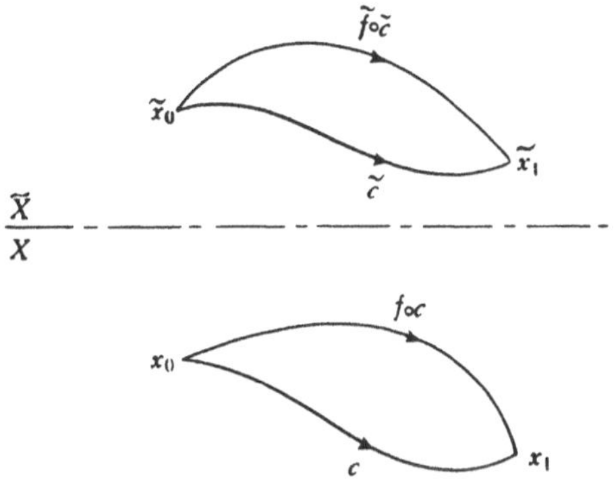

Figure 1

Proof of "$\Rightarrow$".Since $x_1 \in p(\Phi(\tilde{f}))$, there exists $\tilde{x}_1 \in p^{-1}(x_1)$ such that $\tilde{f}(\tilde{x}_1) = \tilde{x}_1$. Take in the connected $\tilde{X}$ a path $\tilde{c}$ from $\tilde{x}_0$ to $\tilde{x}_1$. Then the path $\tilde{f}\circ\tilde{c}$ is also from $\tilde{x}_0$ to $\tilde{x}_1$. Moreover $\tilde{f}\circ\tilde{c} \simeq \tilde{c}$, since $\tilde{X}$ is simply connected. Denote the p-image of $\tilde{c}$ by $c = p\circ\tilde{c}$, a path in X from x_0 to x_1. Then

$$c = p\circ\tilde{c} \simeq p\circ(\tilde{f}\circ\tilde{c}) = (p\circ\tilde{f})\circ\tilde{c} = (f\circ p)\circ\tilde{c} = f\circ c.$$

Proof of "$\Leftarrow$". Denote by $\tilde{c}$ the lifting of the path c which starts from $\tilde{x}_0$. Then the path $\tilde{f}\circ\tilde{c}$ is the lifting of the path $f\circ c$ which starts from $\tilde{x}_0$.By virtue of Theorem B 1.5, $c \simeq f\circ c \Rightarrow \tilde{c}$ and $\tilde{f}\circ\tilde{c}$ have the same terminal point $\tilde{x}_1$. Hence $\tilde{f}(\tilde{x}_1) = \tilde{x}_1$ and $p(\tilde{x}_1) = x_1$; i. e. $x_1 \in p(\Phi(\tilde{f}))$. □

2.2 Corollary. *The identical self-mapping id of any finite connected polyhedron X has exactly one nonempty fixed point class, namely, $\Phi(id) = X$.* □

Remark. Theorem 2.1 is also due to Nielsen ([32 I],p. 289). Its significance is a geometrical characterization of nonempty fixed point class directly in terms of paths in X instead of via the universal covering space $\tilde{X}$, and is often regarded as a definition of fixed point classes, equivalent to Definition 1.5. For convenience, hereafter we refer to Theorem 2.1 and Definition 1.5 as the "*direct*" and *"indirect" definitions* respectively.

Although we have emphasised the "direct" definition in [2] and [36], we prefer the "indirect" definition. With the "indirect" definition it is easier for us to keep in mind that there are both empty and nonempty fixed point classes, and to trace the correspondence between empty and nonempty fixed point classes under a homotopy (see §§ 3—4). After all, nonemptiness or emptiness of a fixed point class of a self-mapping f is not a homotopy invariant, while $R(f)$ (Definition 1.7, Theorem 3.3) and the Nielsen number $N(f)$ (Definition 5.3, Theorem 6.2) are.

The following theorem will show that the fixed point classes of a self-mapping $f: X \to X$ are separated from each other. This theorem has an especially important corollary.

2.3. Theorem. *Every fixed point class* F *of a self-mapping* f *of* X *is an open subset of the fixed point set* $\Phi(f)$ *considered as a subspace of* X. *In other words, for each* F *there exists in* X *an open set* U *such that* $U \cap \Phi(f) = \mathrm{F}$.

Proof. Suppose $\mathrm{F} = p(\Phi(\tilde{f}))$, $\tilde{f}$ being a given lifting of f. We only need to show that for any $x_0 \in p(\Phi(\tilde{f}))$ there exists in X a neighborhood V of x_0 such that all the fixed points of f in V belong to the fixed point class $p(\Phi(\tilde{f}))$, i.e. $V \cap \Phi(f) \subseteq p(\Phi(\tilde{f}))$.

Since $x_0 \in p(\Phi(\tilde{f}))$, there is a fixed point $\tilde{x}_0$ of $\tilde{f}$ such that $p(\tilde{x}_0) = x_0$. Take an admissible neighborhood W of x_0 (cf. Definition

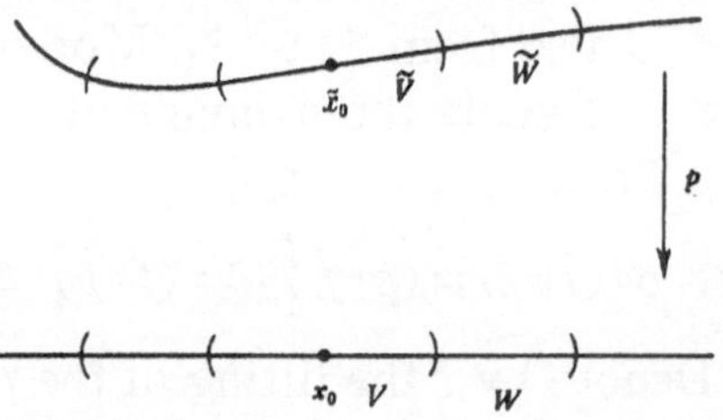

B1.2 of covering spaces), and denote by $\tilde{W}$ the component of $p^{-1}(W)$ which contains $\tilde{x}_0$. Thus $\tilde{W}$ is a neighborhood of $\tilde{x}_0$ in the universal covering space $\tilde{X}$ of X and $p: \tilde{W} \to W$ is a homeomorphism. By virtue of the continuity of $\tilde{f}$ and $\tilde{f}(\tilde{x}_0) = \tilde{x}_0$, there exists in $\tilde{W}$ a neighborhood $\tilde{V}$ of $\tilde{x}_0$ such that $\tilde{f}(\tilde{V}) \subseteq \tilde{W}$. Set $V = p(\tilde{V})$. V is a neighborhood of x_0 in X, since $p: \tilde{W} \to W$ is a homeomorphism.

All these are shown vividly in the accompanying commutative diagram, in which the vertical mappings are all the same homeomorphism p.

$$\begin{array}{ccccc} \tilde{W} & \supseteq & \tilde{V} & \xrightarrow{\tilde{f}} & \tilde{W} \\ \downarrow p & & \downarrow p & & \downarrow p \\ W & \supseteq & V & \xrightarrow{f} & W \end{array}$$

It is easy to show by means of p that the fixed points of $\tilde{f}$ in $\tilde{V}$ and those of f in V are in one-to-one correspondence, and then the conclusion $V \cap \Phi(f) \subseteq p(\Phi(\tilde{f}))$ is arrived at. Finally, the union of all V for all the fixed points of $\mathrm{F} = p(\Phi(\tilde{f}))$ is the open set U required. □

2.4. Corollary. *A self-mapping $\tilde{f}: X \to X$ has only a finite number of nonempty fixed point classes, each of which is a compact subset of X.*

Proof. By virtue of Theorems 1.6 and 2.3, all the fixed point classes of f form an open covering of $\Phi(f)$. Moreover, since $\Phi(f)$ as a subspace of X is compact, this open covering has a finite sub-covering, i. e., f has only a finite number of nonempty fixed point classes.

Every fixed point class is open as well as closed in $\Phi(f)$, and thus is compact in X. □

3. Correspondence between fixed point classes induced by homotopy

Both the "indirect" and the "direct" definitions of the fixed point classes are based on the concept of homotopy. A basic problem is to determine the correspondence induced by a homotopy $H = \{h_t\}: h_0 \simeq h_1$ between the fixed point classes of the self-mappings h_0 and h_1 of our polyhedron X. Let us put together the second part of Definition B 2.3 and Theorem B2.5 or B5.4 as the following.

3.1 Definition. Suppose $H = \{h_t\}$ is a homotopy between the self-mappings h_0 and h_1 of X, $\tilde{X}$ is the universal covering space of X, and $\tilde{h}_0$ is a lifting (Definition 1.1) of h_0. Then there exists a unique lifting $\tilde{H} = \{\tilde{h}_t\}$ of H (the second part of Definition B2.3) such that $\tilde{h}_t$ is a lifting (Definition 1.1) of h_t, for every $t \in I$. Thus the given homotopy H and the given lifting $\tilde{h}_0$ of h_0 determine the unique lifting $\tilde{h}_1$ of h_1. The correspondence from the lifting $\tilde{h}_0$ to the lifting $\tilde{h}_1$ is called the *correspondence induced by the homotopy H* or the *correspondence under the homotopy H,* and is denoted by

$$\tilde{h}_0 \; H \; \tilde{h}_1. \tag{1}$$

Let us investigate more closely the correspondence (1). Firstly, since the homotopy H has the inverse H^{-1} (Definition A3.3), the correspondence (1) has the inverse

$$\tilde{h}_1 \; H^{-1} \; \tilde{h}_0.$$

Hence the correspondence (1) is a one-one correspondence between the liftings of h_0 and those of h_1. Secondly, let $\gamma \in \mathscr{D}$ be any covering motion. Since $\tilde{H}=\{\tilde{h}_t\}:\tilde{h}_0 \simeq \tilde{h}_1 \Rightarrow \gamma \circ \tilde{H} \circ \gamma^{-1}:\gamma \circ \tilde{h}_0 \circ \gamma^{-1} \simeq \gamma \circ \tilde{h}_1 \circ \gamma^{-1}$, the given H and $\tilde{h}_0$ determine *the correspondence from the lifting class* $[\tilde{h}_0]$ *to the lifting class* $[\tilde{h}_1]$:

$$[\tilde{h}_0] H [\tilde{h}_1] \; . \tag{2}$$

Just as (1) is a one-one correspondence, so is (2).

Our "indirect" definition of fixed point classes is based on the lifting classes. Set for convenience $\mathrm{F}_i = p(\Phi(\tilde{h}_i)), i=0,1$. Then (2) gives rise to the one-one correspondence between the labelled fixed point classes (see Notation 1.7):

$$(\mathrm{F}_0, \; [\tilde{h}_0]) H (\mathrm{F}_1, [\tilde{h}_1]). \tag{3}$$

Or, for brevity,

$$\mathrm{F}_0 \, H \; \mathrm{F}_1, \tag{4}$$

which is single-valued but not necessarily one-one.

The discussion above may be stated in the the following two theorems:

3.2 Theorem. *Given a homotopy* $H:f_0 \simeq f_1:X \to X$ *and a fixed point class* F_0 *of* f_0, *a unique fixed point class* F_1 *of* f_1 *is determined, as expressed in* (4). □

3.3 Theorem. *When there is a homotopy* $H:f_0 \simeq f_1:X \to X$, *the Reidemeister numbers* $R(f_0)$ *and* $R(f_1)$ *are equal. In other words, the Reidemeister number of a self-mapping of* X *is a homotopy invariant.* □

Note that for given H and $\tilde{f}_0$, one or both of the fixed point classes F_0 and F_1 in (4) may be empty or nonempty.

Example 3.1. For the circle S^1, the identity mapping $f_0:z \mapsto z$ is homotopic to the self-mapping $f_1:z \mapsto ze^{2\pi\varepsilon i}$, where ε is a small positive constant. $H(z,t) = ze^{2\pi t\varepsilon i}$, $0 \leqslant t \leqslant 1$, is a homotopy from f_0 to f_1.

If the lifting $\tilde{f}_0$ of f_0 is $s \mapsto s+k$, k being an integer, then $\tilde{f}_t(s)$ is $s \mapsto s+k+t\varepsilon$ and $\tilde{f}_1(s)$ is $s \mapsto s+k+\varepsilon$. We have for $k=0$

$$\mathrm{F}_0 = S^1, \; \mathrm{F}_1 = \varnothing,$$

while for $k \neq 0$,

$$\mathrm{F}_0 = \mathrm{F}_1 = \varnothing.$$

Example 3.2. Consider the integral power mapping $z \mapsto z^{-2}$ (see I § 1). Suppose $H(z,t)=z^{-2}$, $H'(z,t)=z^{-2}e^{2\pi ti}$, $0 \leqslant t \leqslant 1$. Obviously, both H and H' are homotopy from f_0 to f_1, where $f_0(z)=f_1(z)=z^{-2}$. Suppose further that the lifting $\tilde{f}_0$ of f_0 is $s \mapsto -2s+k$, k being an integer.

We can find easily that the $\tilde{f}_t$ in $\tilde{H}=\{\tilde{f}_t\}$ is $s \mapsto -2s+k$, while the $\tilde{f}_t'$ in $\tilde{H}'=\{\tilde{f}_t'\}$ is $s \mapsto -2s+k+t$. As shown in Figure I 3, f_0 and f_1 have three nonempty fixed point classes

$$\mathrm{F}_0=\{1\}, \ \mathrm{F}_1=\{e^{\frac{2}{3}\pi i}\}, \ \mathrm{F}_2=\{e^{\frac{4}{3}\pi i}\}.$$

Hence

$$\mathrm{F}_j H \mathrm{F}_j, \ \mathrm{F}_j H' \mathrm{F}_{j+1}, \ j=0,1,2 (\mathrm{mod}\ 3).$$

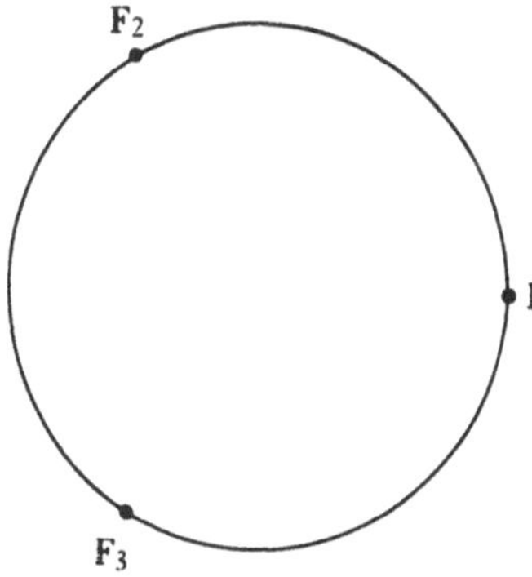

4. Necessary and sufficient condition for the correspondence

Just as we may define a homotopy $H=\{h_t\}: h_0 \simeq h_1$ between self-mappings h_0 and h_1 of our polyhedron X by the single mapping

$$H: X \times I \to X, \ (x,t) \mapsto H(x,t)=h_t(x),$$

in the present section we proceed further to represent this homotopy H by a certain single self-mapping $\mathcal{H}$ of $X \times I$ (cf. [34]). Our aim is to derive first a characterization (Theorem 4.6) of the correspondence (4) in §3, then, for the case when both the fixed point classes in (4) are nonempty, we can derive a necessary and sufficient condition which is a generalization of Theorem 2.1.

4.1 Definition. Suppose $H: X \times I \to X$, $(x,t) \mapsto H(x,t)=h_t(x)$ is a homotopy between self-mappings h_0 and h_1 of X. The following self-mapping of $X \times I$

$$\mathcal{H}: \ X \times I \to X \times I, (x,t) \mapsto \mathcal{H}(x,t)=(H(x,t),t)=(h_t(x),t)$$

is called the *cylindrical representation of H,* and the mapping

$$h_t: X \to X,\ x \mapsto h_t(x)$$

is called the *t-section of H,* for every $t \in I$.

This definition has the following geometrical interpretation (Figure 2). If the restriction of $\mathcal{H}$ on $X \times t$ for a given $t \in I$

$$\mathcal{H} \mid X \times t: X \times t \to X \times t,\ (x,t) \mapsto (h_t(x),t)$$

is taken as *representation* of the t-th section of $\mathcal{H}$, then the heap of all these representations, piled up in order of $t \in I$, is the cylindrical representation $\mathcal{H}$ of H.

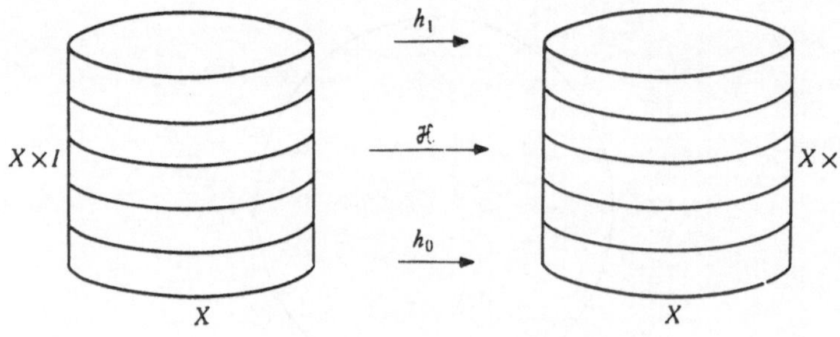

Figure 2

4.2 Lemma. *Suppose $(\tilde{X},p)$ is a covering space of X with the projection p: $\tilde{X} \to X$. Define the mapping*

$$\mathcal{p}: \tilde{X} \times I \to X \times I,\ \mathcal{p}(\tilde{x},t) \mapsto (p(\tilde{x}),t).$$

Then $(\tilde{X} \times I, \mathcal{p})$ is a covering space of $X \times I$ with the projection $\mathcal{p}$. Moreover, $(\tilde{X}, p)$ is the universal covering space of $X \Rightarrow (\tilde{X} \times I, \mathcal{p})$ is also the universal covering space of $X \times I$. □

4.3 Lemma. *Suppose that $(\tilde{X},p)$ is the universal covering space of X, that $H = \{h_t\}: X \times I \to X$ is a homotopy of self-mappings of X, and that $G = \{g_t\}: \tilde{X} \times I \to \tilde{X}$ is a homotopy of self-mappings of $\tilde{X}$. Then G is a lifting of $H \Leftrightarrow$ the cylindrical representation $\mathcal{G}: \tilde{X} \times I \to \tilde{X} \times I$ of G is a lifting of the cylindrical representation $\mathcal{H}: X \times I \to X \times I$ of H.*

Proof. Let us write down first the following diagram:

$$\begin{array}{ccccc} \tilde{X} & \xleftarrow{\ G\ } & \tilde{X}\times I & \xrightarrow{\ \mathscr{G}\ } & \tilde{X}\times I \\ {\scriptstyle p}\downarrow & & \downarrow{\scriptstyle \mathscr{p}} & & \downarrow{\scriptstyle \mathscr{p}} \\ X & \xleftarrow{\ H\ } & X\times I & \xrightarrow{\ \mathscr{H}\ } & X\times I \end{array}$$

The conclusion of our lemma states simply that the commutativity property of the square on the left $\Leftrightarrow$ that on the right. The proof is as follows.

Consider the square on the right of the diagram. By virtue of Definition 4.1 and Lemma 4.2,

$$\begin{aligned} \mathscr{p}\circ\mathscr{G}\,(\tilde{x},t) &= \mathscr{p}\,(G(\tilde{x},t),t) \\ &= \mathscr{p}\,(g_t(\tilde{x}),t) = (p\,(g_t(\tilde{x})),t), \\ \mathscr{H}\circ\mathscr{p}\,(\tilde{x},t) &= \mathscr{H}\,(p\,(\tilde{x}),t) = (H\,(p\,(\tilde{x}),t),t) = (h_t(p\,(\tilde{x})),t). \end{aligned}$$

Therefore $p\,(g_t(\tilde{x})) = h_t(p\,(\tilde{x}))$ for every $t\in I$ (the commutativity property of the square on the left) $\Leftrightarrow \mathscr{p}\circ\mathscr{G} = \mathscr{H}\circ\mathscr{p}$ (the commutativity property of the square on the right). □

4.4 Definition. For any subset A of $X\times I$, the following subset of X

$$[A]_t = \{x\in X : (x,t)\in A\}$$

is called the *t-th section* of A.

4.5 Lemma. *Under the hypothesis of Lemma* 4.3, *for any $t\in I$, we have*

$$[\Phi(\mathscr{H})]_t = \Phi(h_t),$$

$$[\mathscr{p}\,(\Phi(\tilde{\mathscr{H}}))]_t = p\,(\Phi(\tilde{h}_t)),$$

where $\tilde{\mathscr{H}}$ is a lifting of the cylindrical representation $\mathscr{H}$ of $H\,(H\,(x,t) = h_t(x))$.

Proof. It follows obviously from Definition 4.1 that (x,t) is a fixed point of $\mathscr{H} \Leftrightarrow x$ is a fixed point of h_t. Hence we have the first conclusion.

From the definition of $\mathscr{p}$ in Lemma 4.2 and Definition 4.4, it follows that $[\mathscr{p}\,(\tilde{A})]_t = p\,([\tilde{A}]_t)$ for any subset $\tilde{A}$ of $\tilde{X}\times I$; and moreover, just as the first conclusion is valid, so is $[\Phi(\tilde{\mathscr{H}})]_t = \Phi(\tilde{h}_t)$. From these two relations and the first conclusion we find finally our second conclusion on setting $\tilde{A} = \Phi(\tilde{\mathscr{H}})$. □

In terms of the cylindrical representations and their sections, we can characterize the correspondence (4) in § 3 by the following

4.6 Theorem. *Suppose $H=\{h_t\}:X\times I\to X$ is a homotopy of self-mappings of X, and $\mathrm{F}_i=p(\Phi(\tilde{h}_i))$ is a fixed point class of h_i, $i=0,1$. Then $\mathrm{F}_0\,H\,\mathrm{F}_1\Leftrightarrow\mathrm{F}_0$ and F_1 are respectively the 0-th and the 1-th sections of a fixed point class $\not{p}(\Phi(\tilde{\mathcal{H}}))$ of the cylindrical representation $\mathcal{H}$ of H, where $\tilde{\mathcal{H}}$ is a lifting of $\mathcal{H}$ in the sense of Lemmas 4.2 and 4.3.*

Proof. From Definition 3.2, $\mathrm{F}_0 H\,\mathrm{F}_1\Leftrightarrow$there exists a lifting $\tilde{H}$: $\tilde{h}_0\simeq\tilde{h}_1$ such that $\mathrm{F}_i=p(\Phi(\tilde{h}_i))$, $t=0,1$. Then from Lemmas 4.3 and 4.5, $\mathrm{F}_0\,H\,\mathrm{F}_1\Leftrightarrow$there exists a lifting $\tilde{\mathcal{H}}$ of the cylindrical representation $\mathcal{H}$ of H such that $\mathrm{F}_t=[\not{p}(\Phi(\mathcal{H}))]_t$, $t=0,1$. □

Example 4.1. As in Example 3.1, let the self-mappings of S^1 be h_0:$z\mapsto z$, h_1: $z\to ze^{2\pi\varepsilon i}$, and homotopy between them be $H(z,t)=ze^{2\pi t\varepsilon i}$. Then the cylindrical representation of H is

$\mathcal{H}:S^1\times I\to S^1\times I,\ (z,t)\mapsto(ze^{2\pi t\varepsilon i},\ t),$

and the lifting is

$\tilde{\mathcal{H}}:\mathrm{R}^1\times I\to\mathrm{R}^1\times I,(s,t)\mapsto(s+t\varepsilon,t),$

which is the cylindrical representation of the homotopy

$$\tilde{H}:\ \mathrm{R}^1\times I\to\mathrm{R}^1,(s,t)\to s+t\varepsilon,$$

as determined by the lifting $\tilde{h}_0$:$s\mapsto s$ of h_0. The 0-th and 1-th sections of $\not{p}(\Phi(\tilde{\mathcal{H}}))$ are respectively those given in Example 3.1:

$$\mathrm{F}_0=[\not{p}(\Phi(\tilde{\mathcal{H}}))]_0=p(\Phi(\tilde{h}_0))=S^1$$

and

$$\mathrm{F}_1=[\not{p}(\Phi(\tilde{\mathcal{H}}))]_1=p(\Phi(\tilde{h}_1))=\emptyset.$$

Example 4.2. As in Example 3.2, consider the homotopy H:$h_0\simeq h_1\equiv h_0$:$S^1\to S^1$,$(z,t)\mapsto H(z,t)=h_t(z)=z^{-2}e^{2\pi ti}$, and the lifting $\tilde{h}_0$:$s\mapsto-2s+k$, k being an integer. H and $\tilde{h}_0$ determine the lifting $\tilde{H}$: $\mathrm{R}^1\times I\to\mathrm{R}^1$, $(s,t)\mapsto-2s+k+t$. The self-mapping $h_0\equiv h_1$ has three fixed point classes: $\mathrm{F}_0=\{1\}$, $\mathrm{F}_1=\{e^{\frac{2}{3}\pi i}\}$ and $\mathrm{F}_2=\left\{e^{\frac{4}{3}\pi i}\right\}$. For $k=0,1,2$(mod 3), there results $\mathrm{F}_k H\,\mathrm{F}_{k+1}$.

Now let us set $k=0$ and characterize $\mathrm{F}_0\ H\ \mathrm{F}_1$ in terms of the cylindrical representations and sections (Figure 3). The cylindrical representation of H is

$\mathcal{H}$: $S^1\times I\to S^1\times I$, $(z,t)\mapsto\mathcal{H}(z,t)=(h_t(z),t)=(z^{-2}e^{2\pi ti},t)$, and its lifting is (now $k=0$)

$$\tilde{\mathcal{H}}:\mathrm{R}^1\times I\to\mathrm{R}^1\times I,(s,t)\mapsto\mathcal{H}(s,t)=(\tilde{h}_t(s),t)=(-2s+t,t).$$

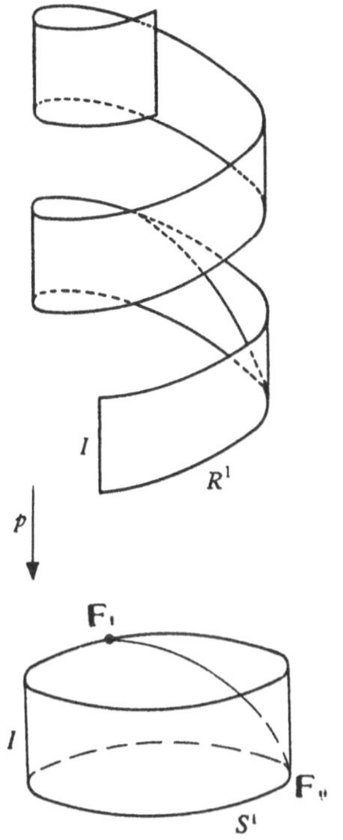

Figure 3

Then

$$p(\Phi(\tilde{N})) = p(\frac{t}{3}, t) = (e^{\frac{2}{3}\pi t i}, t),$$

and finally we have

$$F_0 = [p(\Phi(\tilde{N}))]_0 = \{1\}, F_1 = [p(\Phi(\tilde{N}))]_1 = \{e^{\frac{2}{3}\pi i}\}.$$

Note also the interesting fact that $\Phi(N) = \{(e^{\frac{2}{3}\pi t i}, t) ; t \in I\}$.

These two simple examples illustrate the geometric interpretation of the correspondence (4) in § 3. Moreover, they lead to the following interesting Theorem, a generalization of Theorem 2.1.

4.7 Theorem. *Suppose that $H: h_0 \simeq h_1: X \to X$ is a homotopy, and that x_i is a fixed point in the fixed point class F_i of h_i, $i=0.1$. Then F_0 $HF_1 \Leftrightarrow$ there exists a path c from x_0 to x_1 in X such that* (*Figure* 4)

$$\Delta(H, c) \doteqdot c,$$

where $\Delta(H, c)$ is the diagonal path of H and c (*see Definition* A3.2).

Proof. Let us write down the following chain of three "$\Leftrightarrow$": $F_0 H$ $F_1 \Leftrightarrow$ the points $(x_0, 0)$ and $(x_1, 1)$ in $X \times I$ are in a fixed point class

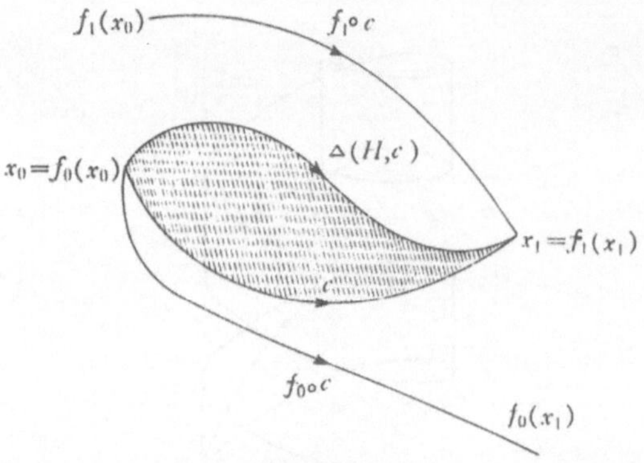

Figure 4

$p(\Phi(\tilde{\mathscr{H}}))$ of the cylindrical representation $\mathscr{H}: X\times I \to X \times I$ of $H \Leftrightarrow$ there exists in $X\times I$ a path $\mathscr{B}$ from $(x_0,0)$ to $(x_1,1)$ such that $\mathscr{H}\circ\mathscr{B} \doteqdot \mathscr{B} \Leftrightarrow$there exists in X a path c from x_0 to x_1 such that $\Delta(H,c) \doteqdot c$. Theorems 4.6 and 2.1 state the validity of the first and the second "$\Leftrightarrow$" respectively. It remains to prove only the third $\Leftrightarrow$.

Proof of the third "$\Rightarrow$". Suppose $\mathscr{B}(t)=(c(t),s(t))$, $t\in I$, is a given path in $X\times I$ from $(x_0,0)$ to $(x_1,1)$. There exists in $X\times I$ a family of paths (see the figure for $\mathscr{B}_\tau(t)$)

$$\mathscr{B}_\tau(t)=(c(t),(1-\tau)s(t)+\tau t),\ \tau\in I.$$

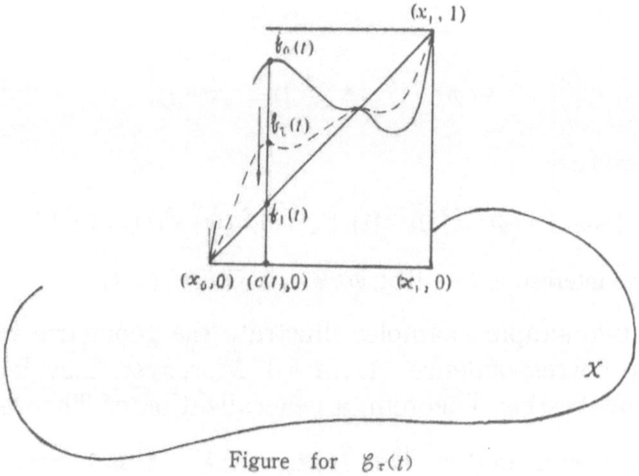

Figure for $\mathscr{B}_\tau(t)$

Since

$$\mathscr{B}_0(t)=\mathscr{B}(t),\ \mathscr{B}_1(t)=(c(t),t),$$

$$\mathscr{B}_\tau(0)=(x_0,0),\ \mathscr{B}_\tau(1)=(x_1,1),$$

there exists therefore in $X\times I$ a homotopy $\mathscr{B}_\tau$ with fixed ends from $\mathscr{B}_0=\mathscr{B}$ to $\mathscr{B}_1$

$$\mathscr{B}_\tau : \mathscr{B} \doteqdot \mathscr{B}_1.$$

From Definition 4.1 and the hypothesis of "⇒", we have

$$\mathscr{H}\circ\mathscr{C}_1 \simeq \mathscr{H}\circ\mathscr{C} \simeq \mathscr{C}.$$

Let $q: X\times I \to X$ be the projection $(x,t)\mapsto q(x,t)=x$. Taking the q-image of $\mathscr{H}\circ\mathscr{C}_1\simeq\mathscr{C}$, we find

$$q\circ\mathscr{H}\circ\mathscr{C}_1 \simeq q\circ\mathscr{C}=c,$$

while

$$\begin{aligned} q\circ\mathscr{H}\circ\mathscr{C}_1 &= q\circ\mathscr{H}(c(t),t)=q(H(c(t),t),t)=H(c(t),t) \\ &= \Delta(H,c)(t). \end{aligned}$$

Hence finally $\Delta(H,c)\simeq c$.

Proof of the third "⇐". From the path c given in the hypothesis, form the path $\mathscr{C}(t)=(c(t),t)$ in $X\times I$. This path is what we require.

□

Example 4.3. By virtue of this theorem, we may obtain directly the conclusion $F_0 H F_1$ in the Example 4.2 by taking the path $c(t)=e^{\frac{2}{3}\pi t i}$ in $X=S^1$, $t\in I$.

Remark. If we set $h_0\equiv h_1$ and take the constant homotopy $H(x,t)\equiv h_0(x)$ in Theorem 4.7, we obtain Theorem 2.1. It is in this sense that Theorem 4.7 is a generalization of Theorem 2.1.

Just as Theorem 2.1 is often regarded as the direct definition of fixed point classes (see Remark in § 2),Theorem 4.7 may be regarded as the direct definition of the correspondence between fixed point classes induced by a homotopy. Historically speaking, this is due to Wecken ([36I] , p.669); and Figure 4 appeared in [2] p. 89.

5. Index of fixed point class. The Nielsen number

So far we have only made use of the theory of universal covering spaces. What we need now is the theory of fixed point index. For convenience we write down the following theorem in a form suitable for our purpose.[1)]

5.1 Theorem. *Suppose X is a connected finite polyhedron, U an open subset of X, and f: $U\to X$ a mapping. If the fixed point set*

---

1) In the original Chinese edition (1979) of our book there was Appendix D containing a simplicial approach to the index theory and a proof of Theorem 5.1. We omit this appendix in the present English version, as the reader has access to an article of similar nature: G. Fournier, A simplicial approach to the fixed point index, pp.73—102 in *Fixed Point Theory*, Lecture Notes in Mathematics, Vol. 886, 1981, Springer.

$\Phi(f)$ of f in U is compact, then an integer $v(f,U)$, called the fixed point index of f in U is defined with the following properties:

(i) **Condition for existence of fixed point.** *If $v(f,U)\neq 0$, then f has at least one fixed point in U.*

(ii) **Homotopy invariance.** *If $F=\{f_t\}$: $f_0\simeq f_1$: $U\to X$ is a homotopy and the union of all the fixed point sets $\Phi(f_t)$*

$$\bigcup_t \Phi(f_t)$$

is compact, then $v(f_0,U)=v(f_1,U)$.

(iii) **Additivity.** *Suppose $U_1,U_2,\cdots,U_s$ are open subsets of U, f has no fixed point in $U-\bigcup_{i=1}^{s} U_i$, and $\Phi(f\mid U_j)$, $j=1,2,\cdots,s$, are pairwise disjoint sets. If $v(f,U)$ is defined, then all $v(f,U_j)$ are defined, and*

$$v(f,U)=\sum_{j=1}^{s} v(f,U_j).$$

(iv) **Normality.** *If f maps U onto a single point of U, then*

$$v(f,U)=1.$$

If f is a self-mapping of X, then

$$v(f,X)=\text{the Lefschetz number } L(f) \text{ of } f,$$

where $$L(f)=\sum_q(-1)^q\operatorname{tr}(f_{q*}),$$

$$f_{q*}:H_q(X,\mathbb{Q})\to H_q(X,\mathbb{Q})$$

is the endomorphism induced by f, and $\operatorname{tr}(f_{q*})$ is the trace of the endomorphism f_{q*} (cf. [3] § 7.2).

(v) **Commutativity.** *Suppose U and V are open subsets of connected finite polyhedra X and Y respectively, and f: $U\to Y$, g: $V\to X$ are mappings. It is easy to see that the composite mappings*

$$g\circ f\colon f^{-1}(V)\to X,\ f\circ g\colon g^{-1}(U)\to Y$$

have respectively the fixed point sets $\Phi(g\circ f\mid f^{-1}(V))(\subseteq U)$ and $\Phi(f\circ g\mid g^{-1}(U))(\subseteq V)$ and that f and g form a pair of mutually inverse homeomorphisms between the sets:

$$\Phi(g\circ f\mid f^{-1}(V))\underset{g}{\overset{f}{\rightleftarrows}}(f\circ g\mid g^{-1}(U)).$$

If $v(g\circ f, f^{-1}(V))$ is defined, then $v(f\circ g,g^{-1}(U))$ is defined as well, and

$$v(g\circ f,f^{-1}(V))=v(f\circ g,\ g^{-1}(U)).$$

(va) **A particular case.** *Suppose X_0 is a subpolyhedron of X, and f: $U \to X$ maps U into X_0. Set $U_0 = U \cap X_0$ and $f_0: U_0 \to X_0$ for the restriction of f. If $v(f, U)$ is defined, then $v(f_0, U_0)$ is also defined and*

$$v(f_0, U_0) = v(f, U).$$

(vi) **Sufficient condition for cancellation of a fixed point.** *Suppose U is homeomorphic to the Euclidien space $\mathbb{R}^k$ of dimension k, f: $U \to X$ has the single fixed point x^* in U, and $v(f, U) = 0$. Then for a neighborhood V of x^* with $\bar{V} \subseteq U$ and $f(\bar{V}) \subseteq U$, there exists a homotopy f_t: $f \simeq g$: $U \to X$ such that g is free from fixed point in U, $g(x) = f(x)$ for all $x \in U - V$, and $g(V) \subseteq U$.* □

When x^* is an isolated fixed point in U of the mapping f: $U \to X$, there is a neighborhood V of x^* such that f has the only fixed point x^* in V. By virtue of the additivity property (iii), $v(f, V)$ is independent of the choice of V. Thus $v(f, V)$ is called the *index of the isolated fixed point* x^* and is denoted also by $v(f, x^*)$. When x^* has a neighborhood homeomorphic to $\mathbb{R}^1$, then the value of $v(f, x^*)$ is just as what was described in Definition I 4.2.

On the basis of this theorem we can proceed to define the Nielsen number, the key concept in the theory of fixed point classes.

Suppose f: $X \to X$ is a self-mapping of a connected finite polyhedron X and $\mathbb{F}$ is a fixed point class of f. From Theorem 2.3 there exists a neighborhood U of $\mathbb{F}$ in X such that $U \cap \Phi(f) = \mathbb{F}$; and then we have $v(f, U)$. If U' is another neighborhood of $\mathbb{F}$ such that $U' \cap \Phi(f) = \mathbb{F}$, then f has no fixed point in both $U - U \cap U'$ and $U' - U \cap U'$. By virtue of Theorem 5.1 (iii),

$$v(f, U) = v(f, U \cap U') = v(f, U').$$

This shows that the index $v(f, U)$ is independent of the choice of such neighborhood U satisfying $U \cap \Phi(f) = \mathbb{F}$, and is a property of $\mathbb{F}$ itself. This gives rise to the following

5.2 Definition. Suppose $\mathbb{F}$ is a fixed point class of the self-mapping f of a connected finite polyhedron X, and U is any neighborhood of $\mathbb{F}$ in X such that $U \cap \Phi(f) = \mathbb{F}$. The index $v(f, U)$ is called the *index of the fixed point class* $\mathbb{F}$ and is denoted by $v(\mathbb{F})$.

5.3 Definition. Suppose f is a self-mapping of a connected finite polyhedron. A fixed point class of f is said to be *non-essential* or *essential* according as its index is zero or nonzero.

5.4 Definition. The number of the essential fixed point classes of a self-mapping f of X is called the *Nielsen number* of f and denoted by $N(f)$. Obviously, $N(f) \leqslant R(f)$.

5.5 Theorem. *The essential fixed point classes of a self-mapping f of X are all nonempty, and empty fixed point classes are all non-essential. There are only a finite number of essential fixed point classes, and hence $N(f)$ is a non-negative integer. f has at least $N(f)$ distinct fixed points.*

Proof. The first statement follows from Theorem 5.1(i). The second follows from the first and Corollary 2.4. The third is obvious. □

Note that a fixed point class may be both nonempty and non-essential.

5.6 Theorem. *The sum of the indices of all essential fixed point classes (or of all nonempty fixed point classes) of a self-mapping f of X is equal to the Lefschetz number $L(f)$ of f.*

Proof. Let the essential fixed point classes of f be $\mathrm{F}_1, \mathrm{F}_2, \cdots, \mathrm{F}_s$, s being an integer. From Theorem 2.3, there exists neighborhood U_j of F_j in X such that $U_j \cap \Phi(f) = \mathrm{F}_j$. From Definition 5.2, $v(\mathrm{F}_j) = v(f, U_j)$. Since $\Phi(f) = \bigcup_j \mathrm{F}_j$, f has no fixed point in $X - \bigcup_j U_j$; moreover, $\Phi(f \mid U_j) = \mathrm{F}_j$, $j = 1, 2, \cdots, s$, are pairwise disjoint. Hence from Theorem 5.1 (iii) and (iv), $\sum_j v(f, U_j) = v(f, X) = L(f)$;

i. e.
$$\sum_j v(\mathrm{F}_j) = L(f).$$ □

For particular cases of these two theorems, see Proposition I 6.1 and Lemma I 8.2.

6. Homotopy invariance: Index of the fixed point class and the Nielsen number

Suppose there is a homotopy $H: f_0 \simeq f_1: X \to X$. A classical result is the homotopy invariance of the Lefschetz number, namely,

$$L(f_0) = L(f_1).$$

From this result and Theorem 5.6, we have immediately the relation:

$$\sum v(\mathrm{F}_{0,j}) = \sum v(\mathrm{F}_{1,j}),$$

where $\mathrm{F}_{i,j}$ runs over all the essential fixed point classes of $f_i, i=0$, 1. A refinement of this relation is given in the following

6.1 Theorem (Homotopy Invariance of Index of Fixed Point Class). *Suppose H: $f_0 \simeq f_1$: $X \to X$ and F_i is a fixed point class of f_i, $i=0.1$. If*

$$\mathrm{F}_0 H \, \mathrm{F}_1$$

(Theorem 3.2), then

$$v(\mathrm{F}_0) = v(\mathrm{F}_1).$$

Proof. $X \times I$ is also a connected finite polyhedron. Consider the cylindrical representation (Figure 5) $\mathcal{H}: X \times I \to X \times I$ of $H = \{f_t\}$. From Theorem 4.6, F_0 and F_1 are the 0-th and 1-th sections of a fixed point class $\mathcal{F} = p(\Phi(\tilde{\mathcal{H}}))$ of $\mathcal{H}$ respectively. In order to prove our theorem, we will first prove $v(\mathcal{F}) = v(\mathrm{F}_s)$, $s=0,1$.

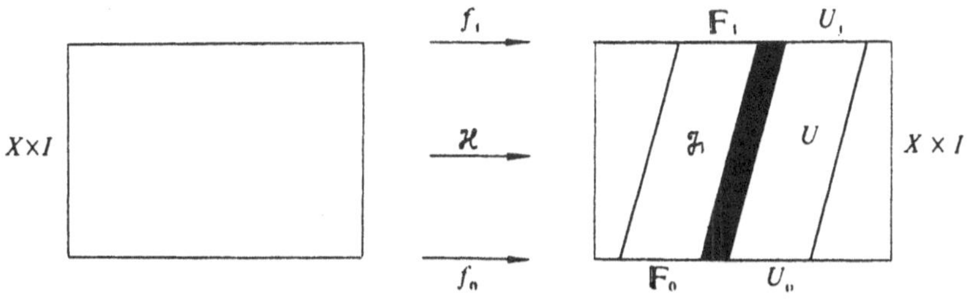

Figure 5

From Theorem 2.3, there is a neighborhood U of $\mathcal{F}$ in $X \times I$ such that $U \cap \Phi(\mathcal{H}) = \mathcal{F}$. The s-th section U_s of U is a neighborhood of F_s evidently, and the s-th section of $\Phi(\mathcal{H})$ is just $\Phi(f_s)$ (Lemma 4.5), $s=0,1$. Hence $U_s \cap \Phi(f_s) = \mathrm{F}_s, s=0,1$. From Definition 5.2, we know $v(\mathcal{F}) = v(\mathcal{H}, U), v(\mathrm{F}_s) = v(f_s, U_s), s=0,1$. In order to complete our proof, we will show that

$$v(\mathcal{H}, U) = v(f_s, U_s), \quad s \in I.$$

Let i_s denote the inclusion of X as the s-th pile in $X \times I$:

$$i_s\colon X \to X \times I,\ x \mapsto i_s(x) = (x,s).$$

Moreover, consider the homotopy, for a given s,

$$G_\tau\colon X \times I \to X \times I,\ (x,t) \mapsto G_\tau(x,t) = (H(x,t), (1-\tau)t + \tau s).$$

Then

$$G_0(x,t) = (H(x,t), t) = \mathcal{H}(x,t),$$

$$G_1(x,t)=(H(x,t),s)=i_s(H(x,t))=i_s\circ H(x,t);$$

and $G_\tau: \mathcal{H}\simeq i_s\circ H$ is a homotopy, and means geometrically that the cylindrical representation $\mathcal{H}$ is pressed onto the s-th thin pile.

It is easy to see that, for any $\tau\in I$, $\Phi(G_\tau)$ is in the compact set $\mathcal{J}$. Then from Theorem 5.1 (ii), for any $s\in I$,

$$v(\mathcal{H},U)=v(i_s\circ H,U).$$

Finally, from H: $X\times I\to X$ and i_s: $X\to X\times I$, there follow $H\circ i_s=f_s$ and $i_s^{-1}(U)=U_s$. On applying Theorem 5.1 (v) (cf. Figure 6) to $H\mid U{:}U\to X$ and i_s: $X\to X\times I$, we find for any $s\in I$,

$$v(i_s\circ H,U)=v(H\circ i_s,\ i_s^{-1}(U))=v(f_s,U_s).$$

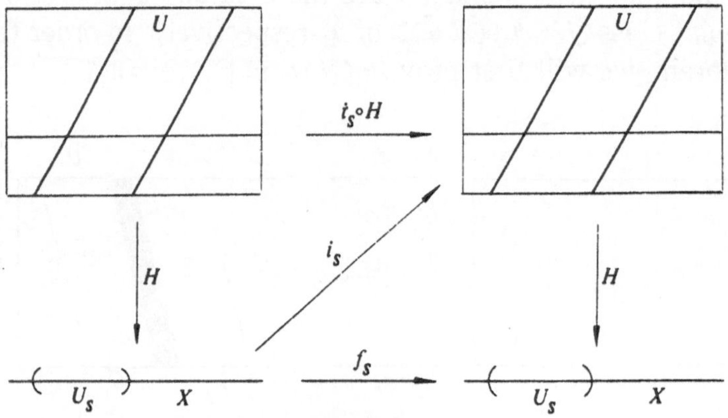

Figure 6

This relation together with that obtained above gives

$$v(\mathcal{H},U)=v(f_s,U_s),\ \text{for any } s\in I;$$

and the particular cases of $s=0,1$ complete our proof. □

For example, in Example 3.2 the fixed point class F_j of the self-mapping f: $S^1\to S^1$, $z\to z^{-2}$ corresponds to F_{j+1}, $j=0,1,2$ (mod 3), under the homotopy $H(z,t)=z^{-2}e^{2\pi ti}$. Hence the indices of the three fixed point classes are all equal, and are equal to $+1$ as shown at the end of I § 4.

From this theorem and Theorem 5.5 there follows

6.2 Theorem (Homotopy Invariance of the Nielsen Number). *Let*

$$H: f_0\simeq f_1{:}X\to X$$

be a homotopy from the self-mapping f_0 to f_1 of a connected finite *polyhedron X which induces the correspondence (Theorem* 3.2)

$$\mathbb{F}_0 H \ \mathbb{F}_1$$

from the fixed point classes of f_0 *to those of* f_1. *Then the correspondence is one-one between the essential fixed point classes of* f_0 *and the essential fixed point classes of* f_1 *and hence*

$$N(f_0) = N(f_1);$$

when $\mathbb{F}_0$ *is non-essential, the corresponding* $\mathbb{F}_1$ *is also non-essential.*

□

6.3 Theorem (The Nielsen Fixed Point Theorem). *Let* $f: X \to X$ *be a self-mapping of a connected finite polyhedron. Then eveny self-mapping homotopic to f has at least* $N(f)$ *fixed points. In other words,*

$$\# \Phi(\langle f \rangle) \geqslant N(f). \qquad \square$$

Example 6.1 the Nielsen number $N(f)$ is defined for any self-mapping f of any finite connected polyhedron, and it remains an open problem to estimate $N(f)$ in general. However,

$$N(f) \leqslant 1$$

for the following two particular cases:

(i) simply connected X and any f,
(ii) any X and $f \simeq id$.

Proof. The conclusion for case (i) follows from Remark 1 after Definition 1.7 and Definition 5.4, and for case (ii) follows from Corollary 2.2 and Theorem 6.2. □

When $N(f) \leqslant 1$, $N(f)$ plays no better role than the Lefschetz number $L(f)$ in the fixed point theory.

For the sake of future need (cf. Theorem IV 3.7) and for justification of Definition I 8.3, let us put down here the following easy consequence of Theorem 6.2, C1.4 and C1.5.

6.4 Corollary. *Let X be a connected finite polyhedron. There exists a positive small number* δ *such that, if two self-mappings f and g of X satisfy the condition:*

$$d(f(x), g(x)) < \delta, \forall x \in X,$$

then

$$N(f) = N(g). \qquad \square$$

Remark. Theorem 6.3 is Theorem N in the first chapter. It gives rise naturally to the problem of evaluation of $N(f)$ and the problem of evaluation of $\#\Phi(\langle f\rangle)$. The two succeeding chapters are devoted respectively to derivation of partial solutions to these two problems.

7. Commutativity: Index of the fixed point class and the Nielsen number

In the applications of the theory of fixed point index the commutativity of the index is sometimes quite useful. We are thus led to ask whether there is also commutativity of the index of fixed point class and of the Nielsen number. We will show in the present section that the answer is positive.

When we set $U=X$ and $V=Y$ in Theorem 5.1 (v), we have the following

7.1 Lemma. *If $f\colon X\to Y$ and $g\colon Y\to X$ are a pair of mappings, then f and g are a pair of mutually inverse homeomorphisms between the fixed point sets $\Phi(g\circ f)(\subseteq X)$ and $\Phi(f\circ g)(\subseteq Y)$.* □

The following lemma is not so obvious as Lemma 7.1 but its proof is immediate.

7.2 Lemma. *If $f\colon X\to Y$ and $g\colon Y\to X$ are a pair of mappings, and x_0,x_1 are two fixed points of $g\circ f$, then x_0, x_1 belong to the same fixed point class of $g\circ f \Leftrightarrow f(x_0)$ and $f(x_1)$ belong to the same fixed point class of $f\circ g$.*

Proof. By virtue of Theorem 2.1, it remains to prove only that there exists in X a path c from x_0 to x_1 such that $(g\circ f)\circ c \fallingdotseq c \Leftrightarrow$ there exists in Y a path d from $f(x_0)$ to $f(x_1)$ such that $(f\circ g)\circ d \fallingdotseq d$. This is obvious, because for a given path c, we can take $d=f\circ c$ and for a given path d we can take $c=g\circ d$. □

7.3 Theorem (Commutativity of Index of Fixed Point Class). *If $f\colon X\to Y$ and $g\colon Y\to X$ are a pair of mappings between connected finite polyhedra, then the f-image of a nonempty fixed point class of $g\circ f\colon X\to X$ is a nonempty fixed point class of $f\circ g\colon Y\to Y$, and f thus establishes a one-one correspondence between the nonempty fixed point classes of $g\circ f$ and those of $f\circ g$. Moreover the indices of corresponding classes are equal.*

Proof. Let F be a nonempty fixed point class of $g\circ f$. From Lemmas 7.1 and 7.2, $f(\mathrm{F})$ is a nonempty fixed point class F' of $f\circ g$, and f: $\mathbb{F} \to \mathbb{F}'$, g: $\mathbb{F}' \to \mathbb{F}$ are mutually inverse homeomorphisms. It remains only to show $v(\mathbb{F}) = v(\mathbb{F}')$.

By virtue of Theorem 2.3, there exists in X a neighborhood U of $\mathbb{F}$ such that $U \cap \Phi(g \circ f) = \mathbb{F}$. Since $g(\mathbb{F}') = \mathbb{F}$, $\mathbb{F}' \subseteq g^{-1}(U)$, i.e. $g^{-1}(U)$ is a neighborhood of $\mathbb{F}'$ in Y. We will show $g^{-1}(U) \cap \Phi(f \circ g) = \mathbb{F}'$. In fact, if a point $y \in g^{-1}(U)$ a fixed point of $f \circ g$, then $g(y) \in U$ is a fixed point of $g \circ f$, i.e.

$$g(y) \in U \cap \Phi(g \circ f) = \mathbb{F},$$

and hence $y = f \circ g(y) \in f(\mathbb{F}) = \mathbb{F}'$.

From Definition 5.2, $v(\mathbb{F}) = v(g \circ f, U)$, $v(\mathbb{F}') = v(f \circ g, g^{-1}(U))$. But for $f \mid U: U \to Y$ and g: $Y \to X$, we have from Theorem 5.1 (v) $v(g \circ f, U) = v(f \circ g, g^{-1}(U))$, and hence $v(\mathbb{F}) = v(\mathbb{F}')$. □

7.4 Theorem (Commutativity of the Nielsen Number). *If f: $X \to Y$ and g: $Y \to X$ are a pair of mappings between connected finite polyhedra, then f establishes a one-one correspondence from the essential fixed point classes of $g \circ f$ to those of $f \circ g$, and consequently*

$$N(g \circ f) = N(f \circ g). \qquad \square$$

Just as the commutativity of fixed point index is useful in applications, so is the commutativity of the Nielsen number, which equates the Nielsen number of a self-mapping of a space X to that of a certain self-mapping of another space Y. Another version of Theorem 7.4 is the following

7.5 Corollary. *When X and Y are connected finite polyhedra, and a self-mapping f of X is the composite mapping of two mappings $h: X \to Y$ and $k: Y \to X$, i.e. $f = k \circ h$, let us write $g = h \circ k$. Then*

$$N(f) = N(g). \qquad \square$$

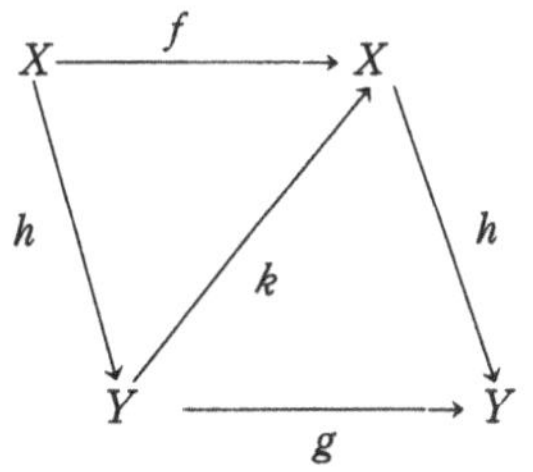

7.6 Corollary[1]. *Suppose f is a self-mapping of a connected finite polyhedron X, X_0 is a subpolyhedron of X, and $f(X) \subseteq X_0$. Then the*

---

1) During our reading the proof-sheets for our earlier Chinese edition, we found Theorem 7.4 and Corollary 7.6 appeared also in [7], p.201.

essential fixed point classes of f are exactly those of the restriction $f|X_0$ of f on X_0, the indices of the corresponding classes are equal, and consequently

$$N(f \mid X_0)=N(f).$$

Proof. In the preceding Corollary, take $Y=X_0$, $h=f$ and k being the inclusion $X_0 \to X$. □

Example 7.1. Let (z,w) represent a generic point of the torus $T^2=S^1\times S^1$, where z and w are both complex numbers with unit norm. Consider the self-mapping f: $T^2\to T^2$, $(z,w)\mapsto (zw,z^{-3}w^{-3})$. Now $f=k\circ h$, where

h: $T^2\to S^1$, $(z,w)\mapsto zw$, and k:$S^1\to T^2$, $z\mapsto(z,z^{-3})$.

Hence $g=h\circ k$:$S^1\to S^1$,$z\mapsto z^{-2}$. Since $N(g)=3$ from Proposition I 8.4, we have from Corollary 7.5

$$N(f)=3.$$

Example 7.2. Let X be a connected finite polyhedron and contain S^1 as a subspace. Consider a self-mapping f of X, whose image $f(X)\subseteq S^1$ and whose restriction $f \mid S^1 \simeq f_{-2}$(see I § 1) on S^1. From Corollary 7.6, $N(f)=N(f \mid S^1)$; and from Proposition I 8.4, $N(f \mid S^1)=3$. Hence $N(f)=3$.

Remark. Up to now our discussion has been restricted to the relation between nonempty fixed point classes of $g\circ f$ and $f\circ g$, because our starting point was Lemma 7.2. If one wishes to discuss the relation between general (empty or nonempty) fixed point classes, one can proceed as follows:

(i) Let $(\tilde{X},p)$ and $(\tilde{Y},q)$ be the universal covering spaces of X and Y respectively. A mapping $\tilde{f}$: $\tilde{X}\to\tilde{Y}$ ($\tilde{g}$: $\tilde{Y}\to\tilde{X}$) is called a lifting of f: $X\to Y$ (g: $Y\to X$) when $q\circ\tilde{f}=f\circ p$ ($p\circ\tilde{g}=g\circ q$).

(ii) Every lifting of $g\circ f$: $X\to X$ decomposes as $\tilde{g}\circ\tilde{f}$, where $\tilde{f}$ and $\tilde{g}$ are liftings of f and g respectively. If $\tilde{f}'$ and $\tilde{g}'$ are also liftings of f and g respectively with $\tilde{g}\circ\tilde{f}=\tilde{g}'\circ\tilde{f}'$, then there is a covering motion $\delta\in\mathscr{D}(\tilde{Y})$ such that $\tilde{f}'=\delta\circ\tilde{f}$ and $\tilde{g}'=\tilde{g}\circ\delta^{-1}$.

(iii) Define a correspondence from the lifting classes of $g\circ f$ to those of $f\circ g$

$$\chi: [\tilde{g}\circ\tilde{f}]\to[\tilde{f}\circ\tilde{g}],$$

which turns out to be single-valued and one-one. Hence χ determines a one-one correspondence between the fixed point classes $p(\Phi(\tilde{g}\circ\tilde{f}))$ of $g\circ f$ and those $q(\Phi(\tilde{f}\circ\tilde{g}))$ of $f\circ g$. This shows the *commutativity of the Reidemeister number*:

$$R(g\circ f)=R(f\circ g).$$

Let us note that there may be no one-one correspondence between the liftings of $g\circ f$ and those of $f\circ g$, for instance when the numbers of leaves of $\tilde{X}$ and $\tilde{Y}$ are different.

(iv) f: $p(\Phi(\tilde{g}\circ\tilde{f}))\to q(\Phi(\tilde{f}\circ\tilde{g}))$ is a homeomorphism between the fixed point classes.

Finally, Theorem 7.3 with all the words "nonempty" deleted can be proved.

The reader might like to carry out as an exercise the procedure outlined above. Although this procedure is not so simple or straight-foreward as Lemma 7.2, it has as by-product the commutativity of the Reidemeister number.

Definition 1.5 and Theorem 2.1 are two aspects of the notion of fixed point classes, and were called in the Remark after Corollary 2.2 the indirect and direct definitions of fixed point classes respectively. In the discussion of the correspondence induced by homotopy and of the commutativity in the present chapter, we have mentioned explicitly which procedure we have followed. An adequate choice of one of the procedures often contributes to an easier and more natural solution to the problem under consideration.

Two topological spaces X and Y are said to be of the *same homotopy type*, denoted by $X\simeq Y$, when there exist mappings $h:X\to Y$ and $k: Y\to X$ such that both $k\circ h\simeq id:X\to X$ and $h\circ k\simeq id: Y\to Y$. Such a pair of mappings (h,k) is called a pair of *homotopy equivalences* for $X\simeq Y$. on the basis of this concept of homotopy type, we introduce (see [27]) the following

7.7 Definition. Let X and Y be two topological spaces of the same homotopy type with homotopy equivalences (h,k). For any self-mapping f of X, let g be a self-mapping of Y such that

$$k\circ f\circ h\simeq g, \text{ or } f\circ h\simeq h\circ g.$$

f and g are said to be the same homotopy type. One can see easily that a pair of homotopy equivalences (h,k) induces in this way a one-one correspondence from the self-mapping classes of X to the self-mapping classes of Y. If a homotopy invariant defined for self-mapping classes of topological spaces takes on the same value as the corresponding classes $\langle f\rangle$ and $\langle g\rangle$, it will then be called a *homotopy type invariant* for the self-mapping classes of the spaces considered.

7.8 Theorem. *The Nielsen number is a homotopy type invariant for self-mapping classes of connected finite polyhedra.*

Proof. Use the notations in Definition 7.7. From Theorems 6.2 and 7.4, we have

$$N(f)=N((k\circ h)\circ f)=N(k\circ(h\circ f))=N((h\circ f)\circ k)=N(g). \qquad \square$$

We leave to the reader to determine whether the Lefschetz number and the Reidemeister number are such homotopy type invariants or not. We shall show in Example IV 3.5 that the least fixed point number $\# \Phi(\langle f \rangle)$ is not such a homotopy type invariant.

Chapter III

EVALUATION OF THE NIELSEN NUMBER

The Lefschetz number $L(f)$ for a self-mapping f of a connected finite polyhedron X can be computed, because it is defined as the alternative sum of the traces of endomorphisms of homology groups (see II 5.1(iv)). Different from $L(f)$, the Nielsen number for f of X is defined in terms of homotopy (see Definition II 5.4), and has been computed only in a few cases.

The first three sections in the present chapter will be devoted to a study of the Reidemeister number $R(f)$, i.e., the number of the fixed point classes of f. In § 1, we introduce the endomorphism $\tilde{f}_\pi$ of $\pi_1(X, x_0)$ by way of a lifting $\tilde{f}$ of f, and then obtain the algebraic formulation of the definition of $R(f)$. In § 2, we obtain a computable lower bound of $R(f)$. In § 3, we investigate the conditions under which $R(f)$ equals this lower bound. In § 4, we define Jiang group $J(f,x_0)$ of Jiang Boju, the key concept in the present chapter. It is a subgroup of the fundamental group $\pi_1(X,f(x_0))$ of X. The two theorems in § 5 determine the value of $N(f)$ when $J(f,x_0)$ is *maximal*, i.e., when $J(f,x_0)=\pi_1(X,f(x_0))$. Each of the three applications in § 6 points out a type of X and f, for which the Jiang group is maximal.

The main results in the present chapter are from [25 I] §§ 3—4. The result in § 3 is due to the author of [25 I] also. Theorem 6.4 is from [2].

So far as we know, the results in §§ 5—6 on the evaluation of $N(f)$ are currently still the most general ones[1].

---

1) Added for this English version. Since the publication of our 1979 Chinese edition, there has been much progress in this theory. Refer to, for instance, the bibliography in the book by Jiang Boju (*Lectures on Nielsen Fixed Point Theory*, Cont. Math.,14, Amer. Math. Soc.) or the forthcoming article by R.F. Brown (Nielsen Fixed Point Theory,in Complete Work of Jakob Nielsen (in English)).

1. Endomorphism $\tilde{f}_\pi$ of $\pi_1(X,x_0)$. $\tilde{f}_\pi$ class. Algebraic definition of $R(f)$

In order to evaluate $N(f)$, it is necessary to have an algebraic formulation of $R(f)$ (Definition II 1.8).

Given the base point $x_0 \in X$, the fundamental group $\pi_1(X,x_0)$ of X can be identified with the group $\mathscr{D}$ of the covering motions of the universal covering space $(\tilde{X},p)$ (Theorem B 3.12). When an arbitrary lifting $\tilde{f}$ in $\tilde{X}$ of a self-mapping f of X is given, any lifting of f in $\tilde{X}$ can be represented uniquely in the form $\alpha\circ\tilde{f}$, $\alpha\in\mathscr{D}=\pi_1(X,x_0)$ (Lemma II 12 (iv)). In this way, a one-one correspondence between $\pi_1(X,x_0)$ and the set of all liftings of f is established in $(\tilde{X},p)$. We will investigate what the image of a lifting class is under this correspondence (see Theorem 1.4).

1.1 Lemma. *Let f be a self-mapping of a connected finite polyhedron X and $\tilde{f}$ a given lifting of f in the universal covering space $\tilde{X}$. For any $\alpha\in\pi_1(X,x_0)$ considered as a covering motion of $\tilde{X}$, a covering motion $\alpha'\in\pi_1(X,x_0)$ is uniquely determined such that the liftings $\alpha'\circ\tilde{f}$ and $\tilde{f}\circ\alpha$ are the same*

$$\alpha'\circ\tilde{f}=\tilde{f}\circ\alpha. \tag{1}$$

The single-valued correspondence $\alpha\mapsto\alpha'$ defined by (1) is an endomorphism of $\pi_1(X,x_0)$, called the endomorphism induced by the given lifting $\tilde{f}$, and denoted by $\tilde{f}_\pi$.

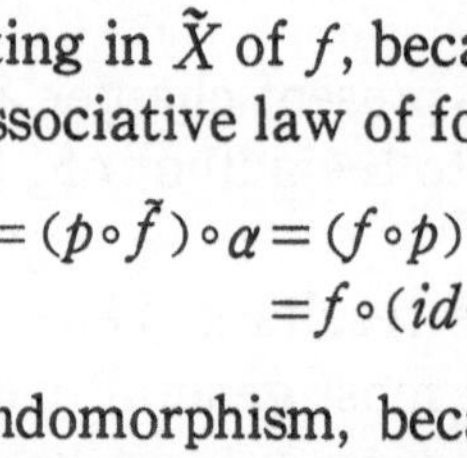

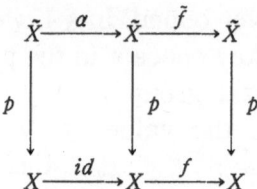

Proof. $\tilde{f}\circ\alpha$ is a lifting in $\tilde{X}$ of f, because (the commutativity of the diagram and the associative law of forming composite functions)

$$\begin{aligned} p\circ(\tilde{f}\circ\alpha)&=(p\circ\tilde{f})\circ\alpha=(f\circ p)\circ\alpha=f\circ(p\circ\alpha)\\ &=f\circ(id\circ p)=f\circ p. \end{aligned}$$

Moreover, $\tilde{f}_\pi$ is an endomorphism, because for $\alpha,\beta\in\pi_1(X,x_0)$ we have

$$\begin{aligned} \tilde{f}_\pi(\alpha\beta)\circ\tilde{f}&=\tilde{f}\circ(\alpha\circ\beta)=(\tilde{f}\circ\alpha)\circ\beta=(\tilde{f}_\pi(\alpha)\circ\tilde{f})\circ\beta\\ &=\tilde{f}_\pi(\alpha)\circ(\tilde{f}\circ\beta)=\tilde{f}_\pi(\alpha)\circ(\tilde{f}_\pi(\beta)\circ\tilde{f}) \end{aligned}$$

$$=(\tilde{f}_\pi(\alpha)\ \tilde{f}_\pi(\beta))\circ\tilde{f}. \qquad \square$$

1.1a Lemma. *Let the lifting $\tilde{f}$ in Lemma 1.1 be defined by $\tilde{x}_0=\langle e_0\rangle\mapsto\tilde{x}_0'=\langle w_0'\rangle$ (see Theorem B5.1). Then the geometrical formula for $\tilde{f}_\pi$ is*

$$\tilde{f}_\pi=w_{0*}'\circ f_\pi,$$

where w_{0*}' and f_π are given in Theorem A5.2 and Definition A5.6 respectively. In other words, for $\alpha=\langle a\rangle$,

$$\tilde{f}_\pi(\alpha)=w_{0*}'\circ f_\pi(\alpha)=\langle w_0'(f\circ a)\,w_0'^{-1}\rangle$$

(see the figure below).

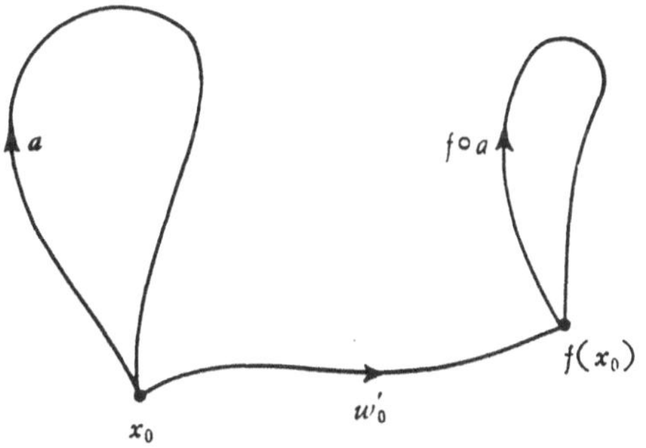

Proof. From Theorem B5.1 and Corollary B5.2, the geometrical formula of $\tilde{f}_\pi$ is derived as follows:

$$\langle c_x\rangle \overset{\alpha}{\to} \langle a\ c_x\rangle \overset{\tilde{f}}{\to} \langle w_0'(f\circ ac_x)\rangle = \langle w_0'(f\circ a)(f\circ c_x)\rangle$$
$$= \langle w_0'(f\circ a)\ w_0'^{-1}\rangle\ \langle w_0'(f\circ c_x)\rangle=(w_{0*}'\circ f_\pi)(\alpha)\circ\tilde{f}\,(\langle c_x\rangle).$$

In other word,

$$\tilde{f}\circ\alpha=\alpha'\circ\tilde{f}=\tilde{f}_\pi(\alpha)\circ\tilde{f}=(w_{0*}'\circ f_\pi)(\alpha)\circ\tilde{f}. \qquad \square$$

Remark 1. By means of this geometrical formula one can give at once another proof that $\tilde{f}_\pi$ is an endomorphism.

A word of warning about the notation $\tilde{f}_\pi$ and f_π is in order here. Unlike f_π (Definition A5.6 with $Y=X$) which is the homomorphism from $\pi_1(X,x_0)$ to $\pi_1(X,f(x_0))$ induced by $f:X\to X$, $\tilde{f}_\pi$ (Lemmas 1.1 and 1.1a) is the endomorphism of $\pi_1(X,x_0)$ induced by $\tilde{f}$: $\tilde{X}\to\tilde{X}$, but not the homomorphism $(\tilde{f})_\pi$: $\pi_1(\tilde{X},x_0)\to\pi_1(\tilde{X},\tilde{f}(\tilde{x}_0))$, where the fundamental group of $\tilde{X}$ is trivial.

1.2 Lemma. *Let f be a self-mapping of X, α, $\alpha'\in\pi_1(X,x_0)$ considered as covering motions of the universal covering space $\tilde{X}$ of*

X, and $\tilde{f}$ be a given lifting of f in $\tilde{X}$. $\alpha\circ\tilde{f}$ and $\alpha'\circ\tilde{f}$ are in the same lifting class of $f \Leftrightarrow$ there exists a covering motion $\gamma\in\pi_1(X,x_0)$ such that in $\pi_1(X,x_0)$

$$\alpha'=\gamma\alpha\tilde{f}_\pi(\gamma^{-1}). \tag{2}$$

Proof. $\alpha'\circ\tilde{f}$ and $\alpha\circ\tilde{f}$ are in the same lifting class (Definition II 1.3) of $f\Leftrightarrow$ there exists an element $\gamma\in\pi_1(X,x_0)$ such that

$$\begin{aligned}\alpha'\circ f&=\gamma\circ(\alpha\circ\tilde{f})\circ\gamma^{-1}=\gamma\alpha\circ(\tilde{f}\circ\gamma^{-1})\\&=\gamma\alpha\tilde{f}_\pi(\gamma^{-1})\circ\tilde{f}.\end{aligned}$$ □

1.3 Definition. Denote the relation between α' and α given in (2) by $\alpha'\sim\alpha$. Obviously this is an equivalence relation in $\pi_1(X,x_0)$, and separates $\pi_1(X,x_0)$ into disjoint equivalent classes. The relation is called the *equivalence relation induced by $\tilde{f}$*, and the classes the *$\tilde{f}_\pi$ classes of $\pi_1(X,x_0)$*. The $\tilde{f}_\pi$ class containing α is the set

$$\{\xi\alpha\tilde{f}_\pi(\xi^{-1}):\forall\xi\in\pi_1(X,x_0)\}.$$

From Lemma 1.2 we have immediately the following

1.4 Theorem. *Let f be a self-mapping of a connected finite polyhedron X, $\tilde{f}$ a given lifting of f in the universal covering space $\tilde{X}$ of X, and $\tilde{f}_\pi$ the endomorphism of $\pi_1(X,x_0)$ induced by $\tilde{f}$. Then the $\tilde{f}_\pi$ classes and the lifting classes of f are in one-one correspondence. The number of $\tilde{f}_\pi$ classes is determined uniquely by f, and is independent of the choice of the lifting $\tilde{f}$.* □

1.5 Corollary. *The number of $\tilde{f}_\pi$ classes determined by f in Theorem 1.4 is equal to the Reidemeister number $R(f)$.* □

This corollary may be regarded as the *algebraic definition* of $R(f)$, while Definition II 1.8 is geometric.

Example 1.1. Suppose f is a self-mapping of the circle S^1 and its mapping degree is n, and $\tilde{f}$ is the lifting of f, determined by a path w_0' from x_0 to $f(x_0)$ in S^1. $\pi_1(S^1,x_0)$ is infinite cyclic, say with generator γ. Then from Lemma 1.1a and Definition 1.3, the geometrical formula of $\tilde{f}_\pi$ and the $\tilde{f}_\pi$ class containing α are respectively

$$\tilde{f}_\pi=w'_{0*}\circ f_\pi:\ \gamma\to\gamma^n,\ \xi=\gamma^k\to\gamma^{nk},\ k \text{ integer},$$
$$\{\xi\alpha\tilde{f}_\pi(\xi^{-1}):\forall\xi\in\pi_1(S^1,x_0)\}=\{\alpha\gamma^{(1-n)k}:\ k \text{ integer}\}.$$

Hence, when $n=1$, every element α of $\pi_1(S^1,x_0)$ is by itself a $\tilde{f}_\pi$ class. When $n\neq1$, there are $|1-n|$ $\tilde{f}_\pi$ classes, namely, the $\tilde{f}_\pi$ classes obtained on setting the following elements

$$\gamma,\ \gamma^2,\cdots,\gamma^{|1-n|},$$

in turn as α. Thus, according as $n=1$ or $\neq1$, $R(f)$ is infinite or $|1-n|$.

Example 1.2. If X is simply connected, then $R(f)=1$ for an arbitrary self-mapping f of X.

Example 1.3. If a self-mapping f of a connected finite polyhedron X is homotopic to the identity mapping of X, i.e. $f \simeq id$, then

$$R(f)=R(id) \geqslant \# \; Z(\pi_1(X,x_0)) \geqslant 1.$$

The notation here needs the following explanation: For any group A, the set of all such elements $\alpha \in A$ which commute with every element of A, i. e. the set

$$\{\alpha \in A: \alpha\xi\alpha^{-1}=\xi, \forall \xi \in A\}$$

is a subgroup of A, called the *center* of A and denoted by $Z(A)$. It can be easily seen that $Z(A)$ is a commutative group. The number of elements of $Z(A)$ is denoted by $\# Z(A)$.

Proof. The first equality follows from Theorem II 3.3. The last "$\geqslant$" holds, since $Z(\pi_1(X,x_0))$ contains at least the identity element.

In order to show the first "$\geqslant$", take the identity mapping $\widetilde{id}$ of $\tilde{X}$ as the lifting $\tilde{f}$ of $f=id$ of X; from Corollary 1.5, $R(id)$ is equal to the number of $\widetilde{id}_\pi$ classes. Moreover, the endomorphism $\widetilde{id}_\pi$ of $\pi_1(X,x_0)$ defined by (1) in Lemma 1.1 for $\tilde{f}=\widetilde{id}$ is the identity $\alpha \mapsto \alpha'=\alpha$. Hence the $\widetilde{id}_\pi$ class containing any given element $\alpha \in \pi_1(X,x_0)$ is

$$\{\xi\alpha\xi^{-1}: \forall \xi \in \pi_1(X,x_0)\}.$$

In particular, when $\alpha \in Z(\pi_1(X_1,x_0))$, the $\widetilde{id}_\pi$ class consists of the single element α. This completes our proof. □

Remark 2. The formula of $\tilde{f}_\pi$ in Example 1.1 is obtained from the formula in Lemma 1.1a, while $\widetilde{id}_\pi$ is shown to be the identity endomorphism by virtue of Lemma 1.1 These two procedures are different; however, when one procedure works, the other also does.

From Corollary 1.5, we learn that only the number $R(f)$ of the $\tilde{f}_\pi$ classes is independent of the choice of $\tilde{f}$. We are going now to point out how the endomorphism $\tilde{f}_\pi$ and the $\tilde{f}_\pi$ classes depend on $\tilde{f}$.

1.6 Theorem. *Let f be a self-mapping of a connected finite polyhedron X and $\tilde{f}$ a given lifting of f in the universal covering space $\tilde{X}$. If $\tilde{f}'$ is another lifting of f and $\tilde{f}'=\beta\circ\tilde{f}$, $\beta \in \pi_1(X,x_0)$ being considered as a covering motion, then $\tilde{f}_\pi$ and $\tilde{f}'_\pi(=(\beta\circ\tilde{f})_\pi)$ satisfy the relation:*

$$\tilde{f}'_\pi(\xi)=\beta\tilde{f}_\pi(\xi)\beta^{-1}, \forall \xi \in \pi_1(X,x_0).$$

Proof. From Lemma 1.1, for any ξ, considered as a covering motion, we have

$$\begin{aligned}\tilde{f}'\circ\xi&=\tilde{f}'_\pi(\xi)\circ\tilde{f}'=\tilde{f}'_\pi(\xi)\circ\beta\circ\tilde{f}\\&=(\tilde{f}'_\pi(\xi)\beta)\circ\tilde{f};\end{aligned}$$

moreover,

$$\begin{aligned}\tilde{f}'\circ\xi&=\beta\circ\tilde{f}\circ\xi=\beta\circ(\tilde{f}\circ\xi)=\beta\circ(\tilde{f}_\pi(\xi)\circ\tilde{f})\\&=(\beta\tilde{f}_\pi(\xi))\circ\tilde{f}.\end{aligned}$$

These two relations give the desired conclusion. □

1.7 Corollary. *Under the hypothesis of Theorem* 1.6, α *and* α' ($\in \pi_1(X,x_0)$) *are in the same* $\tilde{f}'_\pi$ *class* $\Leftrightarrow$ $\alpha\beta$ *and* $\alpha'\beta$ *are in the same* $\tilde{f}_\pi$ *class.*

Proof. From Definition 1.3 and Theorem 1.6, α and α' are in the same $\tilde{f}'_\pi$ class $\Leftrightarrow$ there exists $\gamma\in\pi_1(X,x_0)$ such that $\alpha'=\gamma\alpha\tilde{f}'_\pi(\gamma^{-1})=\gamma\alpha\beta\tilde{f}_\pi(\gamma^{-1})\beta^{-1}$ $\Leftrightarrow$ there exists $\gamma\in\pi_1(X,x_0)$ such that $\alpha'\beta=\gamma(\alpha\beta)\tilde{f}_\pi(\gamma^{-1})$. □

1.8 Corollary. *Under the hypothesis of Theorem* 1.6, $\tilde{f}'=\beta\circ\tilde{f}$ *and* $\tilde{f}$ *are in the same lifting class of f, in other words,* β *and the identity element of* $\pi_1(X,x_0)$ *are in the same* $\tilde{f}_\pi$ *class* $\Rightarrow$ *there exists* $\gamma\in\pi_1(X,x_0)$ *such that*

$$\begin{aligned}\tilde{f}'_\pi(\xi)&=(\gamma\tilde{f}_\pi(\gamma^{-1}))\tilde{f}_\pi(\xi)(\gamma\tilde{f}_\pi(\gamma^{-1}))^{-1}\\&=\gamma\tilde{f}_\pi(\gamma^{-1}\xi\gamma)\gamma^{-1},\ \forall\xi\in\pi_1(X,x_0).\end{aligned}$$

Proof. From Corollary 1.2 and Definition 1.3, the two statements on the left of "$\Rightarrow$" are equivalent. From Corollary 1.2, β and the identity element are in the same $\tilde{f}_\pi$ class $\Rightarrow$ there exists γ such that

$$\beta=\gamma\tilde{f}_\pi(\gamma^{-1}).$$

Then from Theorem 1.6, we obtain the desired conclusion. □

This corollary has no application in the present chapter. It is interesting historically, as shown in the following remark.

Remark 3. Nielsen in his pioneering work ([32 I]) investigated the case of any closed orientable surface X and its self-mapping f. First, in [32 I] §§ 21—22, he considered only the case when f is a homeomorphism and hence its lifting $\tilde{f}$ in the universal covering surface $\tilde{X}$ is also a homeomorphism. In this case, for a given $\alpha\in\pi_1(X,x_0)$ considered as a covering motion, he obtained the corresponding covering motion $\alpha'=\tilde{f}\circ\alpha\circ\tilde{f}^{-1}$ (let the reader prove that α' is a covering motion), and defined the automorphism I of $\pi_1(X,x_0)$:

$$I:\ \alpha\mapsto\alpha'=\tilde{f}\circ\alpha\circ\tilde{f}^{-1},\ \text{or}\ I(\alpha)\circ\tilde{f}=\tilde{f}\circ\alpha. \tag{3}$$

This makes it clear that the expression (1) and the endomorphism $\tilde{f}_\pi$ in our

Lemma 1.1 may be regarded as respective generalizations of the expression (3) and the automorphism I. The totality of I's its generalization is our $\tilde{f}'_\pi$ was called by Nielsen the automorphism family induced by the homeomorphism f.

After this preliminary Nielsen began to consider in § 27 the self-mapping f of the surface X, homotopic to a homeomorphism, and made use of the automorphism family induced by the homeomorphism. Finally in § 37, Nielsen formulated the general problem as follows: For a selfmapping f in the mapping class of a homeomorphism, determine from the automorphism family induced by the homeomorphism the Nielsen number $N(f)$ and the index of each essential fixed point class.

Incidentally, we mention here that our Corollary 1.8 and Theorem 1.4 are the respective generalizations of a formula and the Theorem 13 in [32 I], p.289.

2. A lower bound of $R(f)$

Let us now introduce some general technical terms and notations in the group theory. Let A and B be groups. A homomorphism $\varphi: A \to B$ is called surjective if $\varphi(A)=B$, and injective if $\varphi(A)$ and A are isomorphic, i.e. $\varphi(A) \approx A$.The image $\varphi(A)$ of A is a subgroup of B, denoted by Im φ. The inverse $\varphi^{-1}(e)$ of the identity element e of B is a subgroup of A, called the kernel of φ and denoted by Ker φ. If Im φ is a normal subgroup of B, then the quotient group $\beta/\mathrm{Im}\ \varphi$ is called the cokernel of φ, and is denoted by Coker φ.

2.1 Theorem. *If f is a self-mapping of a connected finite polyhedron X, then*

$$R(f) \geqslant \#\ \mathrm{Coker}\ (id - f_{1*}) \geqslant 1,$$

where id and f_{1*} denote respectively the identity isomorphism and the endomorphism of the 1-dimensional homology group $H_1(X)$ with integral coefficients. If $\pi_1(X, x_0)$ is commutative, then

$$R(f) = \#\ \mathrm{Coker}\ (id - f_{1*}).$$

Proof. 1) From Lemma 1.1a, we have $\tilde{f}_\pi = w'_{0*} \circ f_\pi$. From Theorem A5.8, the following diagram is commutative:

$$\begin{array}{ccc} \pi_1(X,x_0) & \xrightarrow{\tilde{f}_\pi} & \pi_1(X,x_0) \\ \theta \downarrow & & \downarrow \theta \\ H_1(X) & \xrightarrow{f_{1*}} & H_1(X) \end{array}$$

2) From Definition 1.3, any element of the $\tilde{f}_\pi$ class containing α may be expressed in the form

$$\alpha' = \gamma\alpha\tilde{f}_\pi(\gamma^{-1}),\ \gamma \in \pi_1(X,x_0).$$

Then, the commutativity of the above diagram implies

$$\begin{aligned}\theta(\alpha') &= \theta(\gamma\alpha\tilde{f}_\pi(\gamma^{-1})) = \theta(\gamma)+\theta(\alpha)+\theta(\tilde{f}_\pi(\gamma^{-1}))\\ &= \theta(\gamma)+\theta(\alpha)+f_{1*}(\theta(\gamma^{-1}))\\ &= \theta(\alpha)+(id-f_{1*})(\theta(\gamma)).\end{aligned}$$

Therefore, there exists an element $\gamma \in \pi_1(X,x_0)$ such that

$$\theta(\alpha')-\theta(\alpha) = (id-f_{1*})(\theta(\gamma)) \in (id-f_{1*})(H_1(X)).$$

3) Let

$$\eta\colon H_1(X) \to \text{Coker }(id-f_{1*}) \tag{1}$$

be the natural homomorphism. Consider

$$\pi_1(X,x_0) \xrightarrow{\ \theta\ } H_1(X) \xrightarrow{\ \eta\ } \text{Coker }(id-f_{1*}). \tag{2}$$

Now $\eta\circ\theta$ is onto, since both η and θ are; moreover, the $\eta\circ\theta$ images of all elements of a $\tilde{f}_\pi$ class are the same element of Coker $(id-f_{1*})$. This proves the first conclusion of our theorem.

4) When $\pi_1(X,x_0)$ is commutative, θ above is an isomorphism. Then a necessary and sufficient condition for $\alpha' \sim \alpha$ (Definition 1.3) is that there exists $\gamma \in \pi_1(X,x_0)$ such that

$$\theta(\alpha')-\theta(\alpha) = (id-f_{1*})(\theta(\gamma)),$$

or, equivalently,

$$\eta\circ\theta(\alpha')-\eta\circ\theta(\alpha)=0.$$

Thus there follows the second conclusion. □

3. Conditions for $R(f) = \#$ Coker $(id-f_{1*})$

For the equivalence relation $\sim$ in Definition 1.3, there is the following simple rule of operation.

3.1 Lemma. *Let $\tilde{f}$ be a lifting of $f\colon X\to X$ and induce the equivalence relation $\sim$ in $\pi_1(X,x_0)$ (see Definition 1.3). Then, for any $\alpha,\beta \in \pi_1(X,x_0)$, there exist*

$$\alpha\beta \sim \beta\tilde{f}_\pi(\alpha);$$

and, in particular,

$$\alpha \sim \tilde{f}_\pi(\alpha).$$

Proof. From Definition 1.3, we have

$$\beta \tilde{f}_\pi(\alpha) = \alpha^{-1}(\alpha\beta)\tilde{f}_\pi(\alpha) \sim \alpha\beta.$$

For β=the identity $\in \pi_1(X, x_0)$, we have $\alpha \sim \tilde{f}_\pi(\alpha)$. □

The main result in this section is the following Theorem 3.2. It goes further than Theorem 1.4 and states conditions for the $\tilde{f}_\pi$ classes to be independent of the choice of $\tilde{f}$.

3.2 Theorem. *Let $\tilde{f}$ be a lifting of a self-mapping f of a connected finite polyhedron X, and induce the equivalence relation $\sim$ in $\pi_1(X, x_0)$ (see Definition 1.3). Then any two of the following four statements are equivalent:*

(i) *The set of $\tilde{f}_\pi$ classes in $\pi_1(X, x_0)$ is independent of the choice of the lifting $\tilde{f}$, in other words, the set of $\tilde{f}_\pi$ classes is the same as the set of $\tilde{f}'_\pi$ classes where $\tilde{f}'$ is any other lifting of f.*

(ii) *For any $\beta \in \pi_1(X, x_0)$, $\alpha \sim \alpha' \Rightarrow \alpha\beta \sim \alpha'\beta$.*

(iii) *For any α, β, $\gamma \in \pi_1(X, x_0)$, $\alpha\beta\gamma \sim \beta\alpha\gamma$.*

(iv) *The surjective homomorphism $\eta \circ \theta$ in (2) in the proof of Theorem 2.1 induces a one-one correspondence between the set of $\tilde{f}_\pi$ classes and* Coker $(id - f_{1*})$.

Proof. (i)$\Leftrightarrow$(ii): According to Corollary 1.7.

(ii)$\Rightarrow$(iii): According to the second conclusion of Lemma 3.1, we have

$$\alpha\beta \sim \tilde{f}_\pi(\alpha\beta) = \tilde{f}_\pi(\alpha)\tilde{f}(\beta),$$

and

$$\tilde{f}_\pi(\alpha) \sim \alpha.$$

From (ii), the latter relation gives

$$\tilde{f}_\pi(\alpha)\tilde{f}_\pi(\beta) \sim \alpha\tilde{f}_\pi(\beta):$$

and from the first conclusion of Lemma 3.1, we obtain

$$\alpha\beta \sim \beta\alpha.$$

Finally, from (ii) again, we have

$$\alpha\beta\gamma \sim \beta\alpha\gamma.$$

(iii)$\Rightarrow$(iv): It remains to show $\eta \circ \theta(\alpha) = \eta \circ \theta(\alpha') \Rightarrow \alpha \sim \alpha'$, when (iii) holds. The proof is divided into three steps.

1) For any commutator $[\alpha,\beta]=\alpha\beta\alpha^{-1}\beta^{-1}$ and any γ, (iii) $\Rightarrow$

$$[\alpha,\beta]\gamma=\alpha\beta(\alpha^{-1}\beta^{-1}\gamma)\sim\beta\alpha(\alpha^{-1}\beta^{-1}\gamma)=\gamma.$$

2) $\theta(\gamma)=\theta(\gamma')\Rightarrow\gamma\sim\gamma'$. In fact, $\theta(\gamma)=\theta(\gamma')$ means $\gamma'\gamma^{-1}\in\operatorname{Ker}\theta$, i.e., from Theorem A 5.8 there are commutators such that

$$\gamma'=[\alpha_1,\beta_1]\,[\alpha_2,\beta_2]\cdots[\alpha_k,\beta_k]\,\gamma.$$

Then $\gamma'\sim\gamma$, from repeated application of 1).

3) Let $\eta\circ\theta(\alpha)=\eta\circ\theta(\alpha')$ now. From the definition of η (see Relation (1) in the proof of Theorem 2.1), there exists $c\in H_1(X)$ such that

$$\theta(\alpha')-\theta(\alpha)=(id-f_{1*})(c)=c-f_{1*}(c).$$

Let $c=\theta(\gamma),\gamma\in\pi_1(X,x_0)$. From the commutative diagram in the proof of Theorem 2.1, we have $f_{1*}(c)=f_{1*}\circ\theta(\gamma)=\theta\circ\tilde{f}_\pi(\gamma)$, and hence

$$\theta(\alpha')=\theta(\alpha)+\theta(\gamma)-\theta(\tilde{f}_\pi(\gamma))=\theta(\gamma\alpha\tilde{f}_\pi(\gamma^{-1})).$$

From 2), we have finally

$$\alpha'\sim\gamma\alpha\tilde{f}_\pi(\gamma^{-1})\sim\alpha.$$

(iv) $\Rightarrow$ (ii): From Theorem 2.1, $\alpha\sim\alpha'\Rightarrow\eta\circ\theta(\alpha)=\eta\circ\theta(\alpha')$. Since $\eta\circ\theta$ is a homomorphism, we have

$$\eta\circ\theta(\alpha\beta)=\eta\circ\theta(\alpha)+\eta\circ\theta(\beta)=\eta\circ\theta(\alpha')+\eta\circ\theta(\beta)=\eta\circ\theta(\alpha'\beta).$$

Finally, from (iv) we have

$$\alpha\beta\sim\alpha'\beta.$$ □

3.3 Corollary. *If one of the four statements in Theorem* 3.2 *holds, then*

$$R(f)=\#\operatorname{Coker}(id-f_{1*}).$$

Proof. It is immediate from the statement (iv) above. □

We are now to point out some conditions sufficient for the validity of the statement (iii) in Theorem 3.2.

3.4 Definition. Let $f\colon X\to X$ be a self-mapping of a connected finite polyhedron X. We say f has the *central property* if the f_π image of $\pi_1(X,x_0)$ is contained in $Z(\pi_1(X,f(x_0)))$:

$$f_\pi(\pi_1(X,x_0))\subseteq Z(\pi_1(X,f(x_0)))$$

(for the Definition of Z, see Example 1.3). For a positive integer k, let f^k denote the k-th *iterate* $f\circ f\circ\cdots\circ f$ (k times) of f, and f_π^k the homomorphism $(f^k)_\pi = (f_\pi)^k : \pi_1(X, x_0) \to \pi_1(X, f^k(x_0))$ (see Theorem A 5.7(i)). We say f has the *property* $P(k)$ or the *eventually commutative property,* if for some positive integer k

$$f_\pi^k(\alpha\beta) = f_\pi^k(\beta\alpha), \quad \forall \alpha, \beta \in \pi_1(X, x_0).$$

This expression means simply that $f_\pi^k(\pi_1(X, x_0))$ is a commutative subgroup of $\pi_1(X, f^k(x_0))$.

From this definition there follows at once that, for any f, the central property$\Rightarrow f_\pi(\pi_1(X, x_0))$ is a commutative group$\Rightarrow P(1)$, and that $P(k) \Rightarrow P(k+1)$. The latter fact justifies the terminology, the eventually commutative property. Let us note that as a condition on f the central property is stronger than $P(1)$.

Example. A self-mapping $f: X \to X$, for which $P(2) \not\Rightarrow P(1)$, and $P(2) \not\Rightarrow$the central property.

Let C_i, $i = 1, 2, 3$, be three disjoint circles with points $x_i \in C_i$, and $X = C_1 \vee C_2 \vee C_3$ their connected sum with x_1, x_2, x_3 identified, i.e., the quotient space obtained from the union $C_1 \cup C_2 \cup C_3$ by identifying x_i into a single point x_0. Similarly we have $C_2 \vee C_3$ by identifying x_2 and x_3 into x_0. Let $f: X \to X$ be a self-mapping of X such that $f(x_0) = x_0$, and the restrictions $f \mid C_i$: $C_i \to C_{i+1}$, $i = 1$, 2, and $f \mid C_3 : C_3 \to C_3$ have positive mapping degree. Since the centers of $\pi_1(X, x_0)$ and $\pi_1(C_2 \vee C_3, x_0)$ are both trivial and $\pi_1(C_3, x_0)$ is infinite cyclic, f has neither the central property, nor the property $P(1)$, but has the property $P(2)$.

It is important to note that the central property, the property of eventual commutativity and property $P(k)$ are all homotopy properties of f, and are independent of the base point x_0. Moreover, the following are obvious:

The identity self-mapping of X has the central property $\Leftrightarrow$ $\pi_1(X, x_0)$ is commutative $\Rightarrow$ any f has the central property. The latter fact shows how our Definition 3.4 arises as the natural bridge from the second conclusion of Theorem 2.1 to Corollary 3.6 we shall soon obtain.

3.5 Lemma. *If $f: X \to X$ has the property $P(k)$ in Definition* 3.4, *and $\tilde{f}$ is any lifting of f in the universal covering space $\tilde{X}$, then*[1]

$$(\tilde{f}_\pi)^k(\alpha\beta) = (\tilde{f}_\pi)^k(\beta\alpha), \ \forall \alpha, \beta \in \pi_1(X, x_0),$$

where $(\tilde{f}_\pi)^k$ denotes the k-th iterate $\tilde{f}_\pi \circ \tilde{f}_\pi \circ \cdots \circ \tilde{f}_\pi$(k times) of $\tilde{f}_\pi$.

Proof. Let the lifting $\tilde{f}$ be given by $\tilde{x}_0 = \langle e_0 \rangle \mapsto \tilde{x}'_0 = \langle w_0 \rangle$ (Theo-

1) To avoid ambiguity, the notation $(\tilde{f}_\pi)^k$ is used instead of $\tilde{f}_\pi^k$. We point out in Definition 3.4 that $f_\pi^k = (f^k)_\pi = (f_\pi)^k$, and in Remark 1, §1 that $\tilde{f}_\pi$ is not $(\tilde{f})_\pi$.

rem B5.1), where w_0' is a path from x_0 to $f(x_0)$ in X. Let c be any loop at x_0 in X, and $\langle c\rangle=\gamma\in\pi_1(X,x_0)$. On applying Lemma 1.1a repeatedly for k times, we find

$$(\tilde{f}_\pi)^k(\langle c\rangle)=\langle w(f^k\circ c)w^{-1}\rangle=\langle w\rangle\langle f^k\circ C\rangle\langle w^{-1}\rangle$$
$$=w_*\circ(f^k)_\pi(\langle c\rangle)=w_*\circ(f_\pi^k)(\gamma),$$

where w is a path from x_0 to $f^k(x_0)$, composed of the series of paths $w_0', f\circ w_0', \cdots, f^{k-1}\circ w_0'$.

Hence $f_\pi^k(\alpha\beta)=f_\pi^k(\beta\alpha)\Rightarrow(\tilde{f}_\pi)^k(\alpha\beta)=(\tilde{f}_\pi)^k(\beta\alpha)$. □

3.6 Corollary. *If f: $X\to X$ has the property of eventual commutativity, or in particular the central property, then*

$$R(f)=\#\ \mathrm{Coker}\ (id-f_{1*}).$$

Proof. Let $\tilde{f}$ be a lifting of f in the universal covering space $\tilde{X}$. We will show that, under our hypothesis, the statement (iii) in Theorem 3.2 holds. In fact, from the hypothesis and Lemma 3.5, there exists a positive integer k such that $(\tilde{f}_\pi)^k(\alpha\beta)=(\tilde{f}_\pi)^k(\beta\alpha)$. Then from the two conclusions in Lemma 3.1, we find that, $\forall\alpha,\beta,\gamma\in\pi_1(X,x_0)$,

$$\alpha\beta\gamma\sim\tilde{f}_\pi(\alpha\beta\gamma)\sim(\tilde{f}_\pi)^2(\alpha\beta\gamma)\sim\cdots$$
$$\sim(\tilde{f}_\pi)^k(\alpha\beta\gamma)=(\tilde{f}_\pi)^k(\alpha\beta)(\tilde{f}_\pi)^k(\gamma)$$
$$=(\tilde{f}_\pi)^k(\beta\alpha)(\tilde{f}_\pi)^k(\gamma)=(\tilde{f}_\pi)^k(\beta\alpha\gamma)\sim\beta\alpha\gamma.$$

Finally our conclusion follows from Corollary 3.3. □

4. Jiang group and three related lemmas

Let us denote as usual a self-mapping of a connected finite polyhedron X by f. The discussion in the preceding three sections has centered on the number $R(f)$ of the fixed point classes of f, but not yet on the numbers $N(f)$ of the essential fixed point classes. The evaluation of $N(f)$ is in fact a problem too general to be dealt with directly.

On the other hand, we have learned in Chapter I that all the fixed point classes of f have equal indices for the case that X is the circle S^1, and in Chapter II when general X and f the index of a fixed point class of f is a homotopy invariant. We are thus led to ask the following rather specific question. Under a *closed homotopy* f_t:$f\simeq f$ (a homotopy from f to f itself), which pairs of fixed point classes of f can correspond to each other and hence have the same index? In the extreme case when every pair of fixed point classes of f happens to correspond to each other under a suitable closed homotopy, then $N(f)$ can be evaluated as follows: According as $L(f) \neq 0$ or $=0$, We have $N(f) = R(f)$ or $=0$ respectively. This is the idea underlying the discussion of the remaining three sections of the present chapter, and leading to a solution to the specific question.

According to the indirect method (see definition after Corollary II 2.2), the fixed point classes of f are defined by way of the lifting classes of f. We shall devote the present section to the question: Which pairs of the lifting classes of f will correspond to each other under a closed homotopy of f?

Let the self-mapping $\tilde{f}: \tilde{X} \to \tilde{X}$ of the universal covering space $\tilde{X}$ of X be the lifting of f determined by

$$\tilde{x}_0 = \langle e_0 \rangle \mapsto \tilde{x}_0' = \langle w_0' \rangle,$$

where w_0' is a given path from x_0 to $f(x_0)$ in X (Theorem B 5.1). Now any lifting of f has a unique representation $\alpha \circ \tilde{f}, \alpha \in \mathscr{D} = \pi_1(X, x_0)$ (Lemma 1.2(iv)). Suppose that under the given closed homotopy $f_t: f \simeq f$ the lifting $\tilde{f}$ corresponds to the lifting $\alpha \circ \tilde{f}$, in other words, that there is a lifting $\tilde{f}_t$ of the closed homotopy f_t such that $\tilde{f}_0 = \tilde{f}$ and $\tilde{f}_1 = \alpha \circ \tilde{f}$ (see II § 3). First of all, we must observe how the covering motion α is related to the lifting $\tilde{f}$ and the closed homotopy f_t.

From Theorem B5.4,

$$\tilde{f}_1(\tilde{x}_0) = \langle w_0' b \rangle,$$

where b denotes the closed path $b(t) = f_t(x_0)$, $t \in I$, at $f(x_0)$. Then, from Corollary B 5.2,

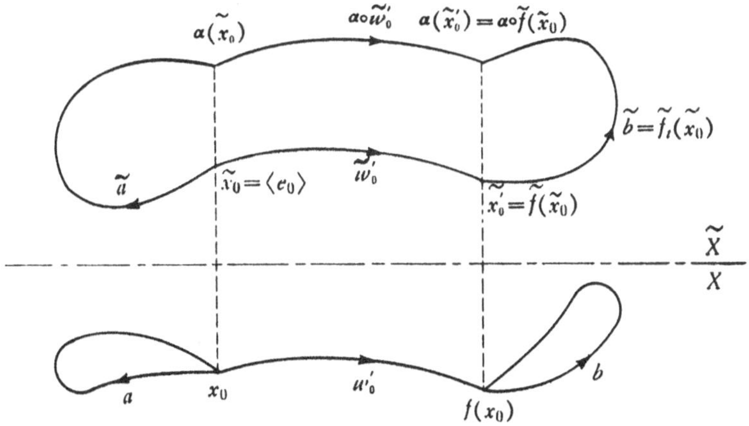

Figure 1

$$\alpha\circ\tilde{f}(\tilde{x}_0)=\alpha(\langle w_0'\rangle)=\alpha\ \langle w_0'\rangle\ .$$

Since $\tilde{f}_1=\alpha\circ\tilde{f}$, from the two expressions just obtained we have (see Figure 1),

$$\langle w_0'b\rangle=\alpha\langle w_0'\rangle\ .$$

Hence from Theorem A 5.2, we have finally

$$\alpha=\langle w_0'bw_0'^{-1}\rangle=w'_{0*}(\langle b\rangle),\tag{1}$$

or

$$\alpha=w'_{0*}(\langle f_t(x_0)\rangle).\tag{2}$$

This consideration leads immediately to the following Definition 4.1 and Lemma 4.2.

4.1 Definition. Let f be a self-mapping of a connected finite polyhedron X. A closed homotopy of f:

$$f_t\colon f\simeq f$$

determines a closed path $f_t(x_0), t\in I$, at $f(x_0)$, where x_0 is the base point of the fundamental group $\pi_1(X,x_0)$. This closed path $f_t(x_0)$ will be called the trace of the base point x_0 under the closed homotopy $f_t\colon f\simeq f$, or simply called the *trace of the closed homotopy* $f_t\colon f\simeq f$. By way of the traces, all closed homotopies of f determine the following subset of $\pi_1(X, f(x_0))$:

$$J(f, x_0)=\{\langle f_t(x_0)\rangle\colon \forall \text{ closed homotopy } f_t\colon f\simeq f\}.$$

Since both the inverse of a closed homotopy and the product of two closed homotopies are also closed homotopies, $J(f,x_0)$ is a subgroup of $\pi_1(X,f(x_0))$, called *Jiang group* for the self-mapping f of X.

Hence, an element β of $\pi_1(X, f(x_0))$ belongs to $J(f,x_0)\Leftrightarrow$ there exists a losed homotopy $f_t\colon f\simeq f$, such that its trace $f_t(x_0)\in\beta$.

Example 4.1. The Jiang group for the identity mapping of the circle S^1 is $J(id, x_0)=\pi_1(S^1,x_0)$.

Proof. Let the point of S^1 be $x=e^{2\pi si}$ (instead of $z=e^{2\pi si}$ used in Chapter I), and $x_0=1$. For $id\colon S^1\to S^1$, $x\mapsto x$, take $f_t(x)=xe^{2\pi ti}=e^{2\pi(s+t)i}$. Clearly f_t: $id\simeq id$ is a closed homotopy of id.

Now, the trace of the closed homotopy is $f_t(x_0)$, a closed path at $x_0=1$, go round S^1 in the positive sense once. It represents a generator γ of $\pi_1(S^1,x_0)$. From the definition of $J(id,x_0)$, $\gamma\in J(id,x_0)$, and hence $J(id,x_0)=\pi_1(S^1,x_0)$.

□

Example 4.2. The Jiang group for a self-mapping f of S^1 is $J(f,x_0)=\pi_1(S^1,f(x_0))$.

Proof. Take $f_t(x)=f(x)e^{2\pi ti}$. Then f_t: $f\simeq f$, and one can proceed as in Example 4.1. □

In terms of Jiang group $J(f,x_0)$, the important relation (2) may be restated as follows

4.2 Lemma. *Let f be a self-mapping of a connected finite polyhedron X, $\tilde{f}$ be the lifting of f determined by*

$$\tilde{x}_0=\langle e_0\rangle \mapsto \tilde{x}_0'=\langle w_0'\rangle$$

(cf. Figure 1), and $\alpha\in\pi_1(X,x_0)$ be regarded as a covering motion of the universal covering space $\tilde{X}$. Then

$$\alpha\in w'_{0*}(J(f,x_0))$$

⇔there exists a closed homotopy f_t: $f\simeq f$ such that, under this homotopy, the lifting $\tilde{f}$ of f corresponds to the lifting $\alpha\circ\tilde{f}$ of f. □

This lemma gives a complete answer to the question which liftings of f can correspond under a closed homotopy of f. Concerning the properties of Jiang groups there are two additional lemmas.

4.3 Lemma[1]. *Between the Jiang groups $J(id,x_0)$ and $J(f,x_0)$ for the identity mapping and the self-mapping f of a connected finite polyhedron X, there are the following relations:*

(i) $J(id,x_0)\subseteq w'_{0*}(J(f,x_0))$, *for any path w_0' from x_0 to $f(x_0)$ in X;*
(ii) $J(id,x_0)=\pi_1(X,x_0)\Rightarrow J(f,x_0)=\pi_1(X,f(x_0))$.

Proof. (i) First, let $\alpha\in J(id,x_0)$. On applying "⇒" in Lemma 4.2 to $f=id$ and $\tilde{f}=\widetilde{id}$ (the identity mapping of $\tilde{X}$, which is the lifting of id determined by the point path $w_0'=e_0$), we know the existence of a closed homotopy

$$f_t: id\simeq id,$$

such that it lifts to

$$\tilde{f}_t: \widetilde{id}\simeq\alpha\circ\widetilde{id}.$$

Secondly, consider a general self-mapping f and its lifting $\tilde{f}$

---

1) For a simpler proof based on the direct method (the Remark in II §2), see [19].

determined by a general path w_0' from x_0 to $f(x_0)$ in X. Now the closed homotopy

$$f_t \circ f: id \circ f \simeq id \circ f, \text{ or } f_t \circ f: f \simeq f$$

has the lifting

$$\tilde{f}_t \circ \tilde{f}: \widetilde{id} \circ \tilde{f} \simeq a \circ \widetilde{id} \circ \tilde{f}, \text{ or } \tilde{f}_t \circ \tilde{f}: \tilde{f} \simeq a \circ \tilde{f}.$$

Since the lifting $\tilde{f}$ is determined by w_0', an application of "⇐" in Lemma 4.2 gives finally $a \in w_{0*}'(J(f,x_0))$. Thus

$$J(id,x_0) \subseteq w_{0*}'(J(f, x_0)).$$

(ii) This follows from (i) since w_{0*}' is an isomorphism.   □

Remark 1. In the proof of (i), we simply made the composition $\tilde{f}_t \circ \tilde{f}$ of $\tilde{f}$ and the homotopy $\tilde{f}_t$. Let us now inverse the order and make the composition $\tilde{f} \circ \tilde{f}_t$. Then

$$\tilde{f} \circ \tilde{f}_t: \tilde{f} = \tilde{f} \circ \widetilde{id} \simeq \tilde{f} \circ a \circ \widetilde{id} = \tilde{f}_\pi(a) \circ \tilde{f}.$$

Since a is any element of $J(id,x_0)$ and $\tilde{f}$ is the lifting of f determined by w_0', by "⇐" in Lemma 4.2, the above relation means just

$$\tilde{f}_\pi(J(id,x_0)) \subseteq w_{0*}'(J(f, x_0)).$$

From Lemma 1.1a and because w_{0*}' is an isomorphism, we obtain the following interesting result:

$$f_\pi(J(id,x_0)) \subseteq J(f, x_0).$$

In [16] the relation (i) is described vividly by saying that $J(id,x_0)$ is *smaller than* $J(f, x_0)$.

Example 4.3. The aim of this example is to illuminate the first step of the proof (i) in Lemma 4.3 with the particular case $X = S^1$. Let us use the notations in Example 4.1. Let $f_t(x) = xe^{2\pi ti}: id \simeq id, f_t(x_0)$ represents the generator γ of $\pi_1(S^1,x_0)$.

Consider the identity $\widetilde{id}: R^1 \to R^1, s \mapsto s$, a lifting of $id: S^1 \to S^1$. The lifting of f_t, determined by $\widetilde{id}$, is

$$\tilde{f}_t(s) = s + t,$$

because the projection $p: R^1 \to S^1$ is the exponenntial mapping (see I § 2) and $p(\tilde{f}_t(s)) = xe^{2\pi ti} = f_t(x)$. Hence

$$\tilde{f}_t: \widetilde{id} \simeq \gamma \circ \widetilde{id}.$$

Example 4.4. The intent of this example is to illustrate the second step of the proof of (i) in Lemma 4.3 with the particular case $X=S^1$. Continue to use the notation in the preceding example. Now suppose that f is an arbitrary self-mapping of S^1, and that $f_t\ (x)=f(x)e^{2\pi ti}$: $f\simeq f$. If $\tilde{f}$ is any lifting of f, then

$$p(\tilde{f}(s))=f(p(s))=f(e^{2\pi si})=f(x).$$

We say that the lifting of $f_t(x)$, determined by $\tilde{f}$, is

$$\tilde{f}_t(s)=\tilde{f}(s)+t,$$

because $p(\tilde{f}_t(s))=f_t(x)$. Hence

$$\tilde{f}_t\colon \tilde{f}\simeq\gamma\circ\tilde{f}.$$

We stated previously the definition of the center $Z(A)$ of a group A. We need in the following a generalization of this definition. Suppose a group B is a subgroup of a group A. The set of all those elements of A which commute with every element of B, i.e.

$$\{a\in A\colon\ a\xi a^{-1}=\xi, \forall\,\xi\in B\},$$

forms a subgroup of A, denoted by $Z(B,A)$ and called the *centralizer of B in A*. It is easy to see that $Z(A,A)=Z(A)$ is commutative, and that for $C\subseteq Z(A), Z(C,A)=A$ is not necessarily commutative.

4.4 Lemma. (i) *Every element of $J(f,x_0)$ and every element of $f_\pi\ (\pi_1(X,x_0))$ are commutative in $\pi_1\ (X,f(x_0))$; in other words,*

$$J(f,\ x_0)\subseteq Z(f_\pi(\pi_1(X,x_0)),\ \pi_1(X,f(x_0)));$$

(ii) $J(id,\ x_0)\subseteq Z(\pi_1(X,x_0))$.

Proof. (i) Let $\tilde{f}$ be the lifting of f determined by

$$\tilde{x}_0=\langle e_0\rangle\rightarrow\tilde{x}_0'=\langle w_0'\rangle,$$

where w_0' is a path from x_0 to $f\ (x_0)$ in X. Then, from the geometrical formula in Lemma 1.1a and the isomorphism w_{0*}' (i) is equivalent to the following

(i′) $w_{0*}'(J(f,\ x_0))\subseteq Z(\tilde{f}_\pi(\pi_1(X,x_0)),\ \pi_1(X,x_0))$.

In other words, every element of $w_{0*}'\ (J(f,x_0))$ and every element of $\tilde{f}_\pi(\pi_1(X,x_0))$ are commutative in $\pi_1(X,x_0)$. We proceed to prove (i′).

Let $a\in w_{0*}'(J(f,x_0))$ be regarded as a covering motion of $\tilde{X}$. We will show that a and every element of $\tilde{f}_\pi(\pi_1(X,x_0))$ are commutative. From Lemma 4.2, there exists a closed homotopy $f_t\colon f\simeq f$ such that for the lifting $\tilde{f}_t$ of f_t determined by $\tilde{f}$,

$$\tilde{f}_t\colon \tilde{f}\simeq a\circ\tilde{f}.$$

Furthermore, for any $\gamma \in \pi_1(X,x_0)$, regarded as a covering motion, we have the homotopy:

$$\begin{aligned}\tilde{f}_t\circ\gamma\,(\text{denoted as } \tilde{f}'_t)\colon \tilde{f}\circ\gamma &\simeq (\alpha\circ\tilde{f})\circ\gamma\\ &=\alpha\circ(\tilde{f}\circ\gamma)=\alpha\circ(\tilde{f}_\pi(\gamma)\circ\tilde{f})\\ &=(\alpha\tilde{f}_\pi(\gamma))\circ\tilde{f},\end{aligned} \tag{1}$$

where the equalities are based on Lemma 1.1. Similarly, we have

$$\begin{aligned}\tilde{f}_\pi(\gamma)\circ\tilde{f}_t\,(\text{denoted as } \tilde{f}''_t)\colon \tilde{f}_\pi(\gamma)\circ\tilde{f}&\simeq\tilde{f}_\pi(\gamma)\circ(\alpha\circ\tilde{f})\\ =(\tilde{f}_\pi(\gamma)\circ\alpha)\circ\tilde{f}&=(\tilde{f}_\pi(\gamma)\alpha)\circ f.\end{aligned} \tag{2}$$

We will verify (cf. Theorem A 1.5) from (1) and (2) the following

$$\alpha\tilde{f}_\pi(\gamma)=\tilde{f}_\pi(\gamma)\,\alpha, \tag{3}$$

which is exactly (i'). The derivation follows. First, since γ and $\tilde{f}_\pi(\gamma)$ are regarded as covering motions, it is easy to verify that both $\tilde{f}'_t$ and $\tilde{f}''_t$ are liftings of f_t. Second, $\tilde{f}'_0=\tilde{f}\circ\gamma$ and $\tilde{f}''_0=\tilde{f}_\pi(\gamma)\circ\tilde{f}$ are the same lifting (Lemma 1.1). Finally, from Theorem B5.4 we have $\tilde{f}'_t=\tilde{f}''_t$, and particularly

$$\tilde{f}'_1=\tilde{f}''_1.$$

From this expression we obtain (3) (cf. Lemma II 1.2 (iv)).

(ii) When f is the identity mapping id, f_π is the identity isomorphism of $\pi_1(X,x_0)$. Then (ii) follows from (1) and $Z(A,A)=Z(A)$. □

Remark 2. In the proof of Lemma 4.4 (i), we have $\tilde{f}_t\circ\gamma=\tilde{f}_\pi(\gamma)\circ\tilde{f}_t$. Noticing $\tilde{f}_t\circ\gamma=(\tilde{f}_t)_\pi(\gamma)\circ\tilde{f}_t$, we obtain also

$$(\tilde{f}_t)_\pi=\tilde{f}_\pi.$$

Making use of the latter of the two procedures mentioned in Remark 2 coming after Example 1.3, we easily obtain the above formula for $\tilde{f}_t\colon s\to s+t$ in Example 4.3 and for $\tilde{f}_t\colon s\to\tilde{f}(s)+t$ in Example 4.4.

5. Evaluation of the Nielsen number in the case of maximal Jiang group

At the beginning of § 4, we proposed a specific question: Given a self-mapping f of a connected finite polyhedron X, which of the fixed point classes of f can correspond to each other under closed homotopies of f and hence have the same index? Lemma 4.2 provides a complete answer to this question. On the basis of the three lemmas in § 4, we now easily obtain the following two fundamental theorems on the evaluation of the Nielsen number $N(f)$ in the extreme case that the Jiang group is maximal.

5.1 Theorem. *Suppose that the Jiang group for a self-mapping f of a connected finite polyhedron X is maximal, i.e.*

$$J(f, x_0) = \pi_1(X, f(x_0)).$$

Then

(i) *f has the central property and hence* $R(f) = \#$ Coker $(id - f_{1*}) \geqslant 1$.

(iia) *When $L(f) \neq 0$, $N(f) = R(f) \geqslant 1$ and the index of each fixed point class of f is $L(f)/N(f) \neq 0$.*

(iib) *When $L(f) = 0$, $N(f) = 0$.*

Proof. (i) Our hypothesis together with Lemma 4.4 (i) gives

$$f_\pi(\pi_1(X,x_0)) \subseteq Z(\pi_1(X,f(x_0))),$$

i.e., f has the central property. Then according to Corollary 3.6, we have $R(f) = \#\ \mathrm{Coker}(1 - f_{1*}) \geqslant 1$.

(ii) Our hypothesis together with Lemma 4.2 implies that, given any two liftings of f, there exists a closed homotopy $f_t: f \simeq f$ of f, under which the two liftings correspond to each other. Hence the indices of all fixed point classes of f are equal to the same integer v. Then $vR(f) = L(f)$.

(iia) When $L(f) \neq 0$, $v \neq 0$. Hence the conclusion.

(iib) When $L(f) = 0$, $v = 0$. Hence every fixed point class is non-essential and $N(f) = 0$. □

5.2 Theorem. *Suppose that the Jiang group for the identity mapping of a connected finite polyhedron X is maximal, i.e.*

$$J(id, x_0) = \pi_1(X, x_0).$$

Then

(i) *$\pi_1(X, x_0)$ is a commutative group, and*

(ii) *for any self-mapping f, $J(f, x_0) = \pi_1(X, f(x_0))$, and the conclusions in Theorem* 5.1 *also hold.*

Proof. (i) Our hypothesis together with Lemma 4.4 (ii) gives $\pi_1(X, x_0) \subseteq Z(\pi_1(X, x_0))$, i.e., $\pi_1(X_1, x_0)$ is a commutative group.

(ii) From our hypothesis, Lemma 4.3 (ii) and Theorem 5.1, we have $J(f, x_0) = \pi_1(X_1 f(x_0))$. □

Example 5.1. Constant mapping c: $X \to x_0 (\in X)$.

The mapping c has the point x_0 as the only fixed point and hence only one nonempty fixed point class. Since the index $v(c,x_0)$ of the fixed point is $+1$ (Theorem II 5.1 (iv)), we have $N(c)=1$.

This result can be obtained also from Theorem 5.1. First we show that $J(c, x_0)=\pi_1(X,x_0)$. Let $\alpha=\langle a\rangle\in\pi_1(X,x_0)$, where a denotes any closed path $a(t)$, $t\in I$, in X at x_0, i.e. with $a(0)=a(1)=x_0$. Set

$$f_t(x)=a(t),\ \forall x\in X.$$

Then $f_t:c\simeq c$ is a closed homotopy of the mapping c, and the trace $f_t(x_0), t\in I$, of the closed homotopy is just the given path a. Hence $\alpha\in J(c,x_0)$ and $J(c, x_0)=\pi_1(X,x_0)$.

Moreover, $c_\pi(\pi_1(X,x_0))$ is the identity element $\langle\{x_0\}\rangle$ of $\pi_1(X,x_0)$, and hence c_{1*} is the null endormorphism of $H_1(X)$. This gives $R(c)=\#\,\mathrm{Coker}\,(id-c_{1*})=1$. From the obvious fact $L(c)=1$, we find $N(c)=1$ from Theorem 5.1.

Example 5.2. Self-mapping f of the circle with mapping degree n. From Example 4.2 and Theorem 5.1 (or Example 4.1 and Theorem 5.2), $L(f)=1-n$, and $N(f)=|1-n|$ (cf. Example 1.1). Then according as $n<1, >1$ or $=1$, we have $v=+1, -1$, or 0 respectively.

Of course, this result can also be obtained without Theorems 5.1 and 5.2.

Example 5.3. Self-mapping f of $L(m;q_1,\cdots,q_n)$, the lens space of dimension $2n+1$.

Definition of $L(m;q_1,\cdots,q_n)$. Let S^{2n+1} denote the unit sphere in R^{2n+2}, given in terms of $n+1$ complex coordinates

$$z_0=r_0e^{i\theta_0},\ \cdots,\ z_n=r_ne^{i\theta_n}$$

by the equation

$$r_0^2+r_1^2+\cdots+r_n^2=1.$$

Let $m\geqslant 2$ be a given integer, and let $q_1,\ \cdots,\ q_n$ be n given natural numbers, relatively prime to m. The rotation

$$\gamma:(z_0,\cdots,z_j,\cdots,z_n)\mapsto(z_0e^{2\pi i/m},\cdots,\ z_je^{2\pi iq_j/m},\cdots,z_ne^{2\pi iq_n/m})$$

generates a cyclic group of order m, a transformation group acting on S^{2n+1} which will be denoted by π_1. Every element of π_1, if not the identity element, has no fixed point. On identifying every two points of S^{2n+1} which correspond under an element of π_1, there results the lens space $L(m;\ q_1,\cdots,q_n)$, or briefly denoted by L. L is a $(2n+1)$- dimensional orientable manifold with S^{2n+1} as the m-leaved (see Corollary B1.6) universal covering space; π_1 is the fundamental group $\pi_1(L,x_0)$ of L, also the group of covering motions of S^{2n+1} (see [11], for $n=1$).

Evaluation of $N(f)$. 1) To show $J(id,x_0)=\pi_1(L,x_0)$. When we consider the identity mapping $\widetilde{id}$ of S^{2n+1} as the lifting of the identity mapping id of L, and the following homotopy

$$\gamma_t: (z_0, \cdots, z_j, \cdots, z_m) \mapsto (z_0 e^{2\pi ti/m}, \cdots, z_j e^{2\pi tiq_j/m}, \cdots, z_n e^{2\pi tiq_n/m}),$$

we find

$$\gamma_t: \widetilde{id} \simeq \gamma \circ \widetilde{id}.$$

Then, from Lemma 4.2,) $\gamma \in J(id, x_0)$. (Please note: in Examples 4.1, 4.2 and 5.1 only the definition of $J(f, x_0)$ was used, but not Lemma 4.2.) Hence $J(id, x_0) = \pi_1(L, x_0)$, and Theorem 5.2 can be applied.

2) To compute $R(f)$ by means of Corollary 3.3. Suppose that by a lifting $\tilde{f}$ of f is induced the following endomorphism of $\pi_1(L, x_0)$

$$\tilde{f}_\pi: \gamma \mapsto \gamma^s, \ 0 \leqslant s \leqslant m.$$

Since both $\pi_1\ (L, x_0)$ and $H_1(L)$ are cyclic groups of order m ($\approx J_m$, the group of integers, mod m), let us denote the generator $\theta(\gamma)$ of $H_1(L)$ by θ briefly. Then

$$f_{1*}: \theta \mapsto s\theta;$$
$$id - f_{1*}: \theta \mapsto (1-s)\theta, \ x\theta \mapsto (1-s)x\theta,$$

where x is an integer mod m. From a theorem in number theory, the congruence

$$(1-s)x \equiv 0 \pmod{m}$$

has exactly $(1-s, m)$ (the greatest common factor of $1-s$ and m, a positive integer) solutions. Hence

$$\#(id - f_{1*})H_1(L) = m/(1-s, m), R(f) = (1-s, m).$$

3) To compute $L(f)$. Since $L(f) = 1 - \deg f$, and since a theorem of Olum ([33] p. 467) gives

$$\deg f = s^{n+1} + km, \ k \text{ an integer},$$

as a consequence of $\tilde{f}_\pi: \gamma \mapsto \gamma^s$, $0 \leqslant s < m$, we find

$$L(f) = 1 - s^{n+1} - km.$$

4) Finally, on applying Theorem 5.1, we have the result: according as $L(f) = 1 - s^{n+1} - km \neq 0$ or $= 0$, then $N(f) = (1-s, m)$ or $= 0$ respectively (cf. [22] p.77, for $n=1$).

6. Applications of Theorems 5.1 and 5.2

If for a topological space X and one of its points e, there exists a mapping

$$\mu: X \times X \longrightarrow X$$

such that

$$\mu(e,x)=\mu(x,e)=x, \forall x \in X,$$

then the space X is called an *H-space,* the point e its *unity point,* and μ its *multiplication.* It is a generalization of Lie and topological groups.

6.1 Theorem. *If a connected finite polyhedron X is an H-space, then*

$$J(id,e)=\pi_1(X,e);$$

and the conclusions of Theorem 5.2 *hold for X and every self-mapping f of X.*

Proof. Take the point e as the base point of X. Let a denote any closed path $a(t)$, $t \in I$, in X at e. By means of the multiplication, we can define the following closed homotopy of the identity mapping id of X:

$$f_t(x)=\mu(a(t),x):\ id \simeq id.$$

It is obvious that the trace $f_t(e)=a(t)$. Hence from definition of $J(id,e)$, we have $\langle a \rangle \in J(id,e)$, where a is any closed path in X at e. □

6.2 Theorem. *If a connected finite polyhedron X is the quotient group of a topological group to an arcwise connected closed subgroup, then*

$$J(id,x_0)=\pi_1(X,x_0),$$

and the conclusions of Theorem 5.2 *hold for X and every self-mapping f of X.*

We omit here the proof and refer the reader to [25 I]. □

A topological space is said to be *non-spherical,* if its homotopy groups $\pi_n(X,x_0)$, $n>1$, are all null.

6.3 Lemma. *If a connected finite polyhedron X is non-spherical, then, for any self-mapping f of X, we have*

$$J(f,x_0)=Z(f_\pi(\pi_1(X,x_0)),\ \pi_1(X,f(x_0))),$$

and in particular,

$$J(id,x_0)=Z(\pi_1(X,x_0)).$$

Proof. By virtue of Lemma 4.4(i), it remains only to prove

$$J(f,x_0)\supseteq Z(f_\pi(\pi_1(X,x_0)),\ \pi_1(X,f(x_0))). \tag{1}$$

For this proof, see [2] pp. 102—103. □

6.4 Theorem. *If a connected finite polyhedron X is non-spherical and its self-mapping f has the central property, then*

$$J(f, x_0) = \pi_1(X, f(x_0)), \tag{2}$$

and the conclusions of Theorem 5.1 *hold for X and f.*

Proof. X is non-spherical$\Rightarrow$(1) for any self-mapping f of X, f has the central property$\Rightarrow$

$$f_\pi(\pi_1(X,x_0)) \subseteq Z(\pi_1(X,f(x_0))).$$

Recalling the fact (the Remark before Lemma 4.4) $C \subseteq Z(A) \Rightarrow Z(C, A) = A$, we obtain (2) from Lemma 6.3. □

Chapter IV

NIELSEN NUMBER AND THE LEAST NUMBER OF FIXED POINTS

Let K be a connected finite simplicial complex and f a self-mapping of the polyhedron $|K|$. In the Introduction of Chapter I, we defined $\# \Phi(\langle f\rangle)$, the least number of fixed points of the mapping class $\langle f\rangle$ of f. Then in Chapter II, the concept of the Nielsen number $N(f)$ was introduced and Nielsen fixed point Theorem II 6.3 was proved. Right after the proof of Theorem II 6.3, we proposed the evaluation problem of $N(f)$ and of $\# \Phi(\langle f\rangle)$. Just as the preceding chapter was devoted to some solutions to the evaluation problem of the first number, this present chapter will be devoted to some solutions to the evaluation problem of the second number.

The main results in the present chapter are as follows. 1) Theorem 2.4 is Theorem 2 in [36 III], a generalization of [1], XIV § 4; 2) Theorem 3.5 is the Theorem in [35a] and gives a combinatorial method of calculating $\# \Phi(\langle id\rangle)$ for general K (Definition 3.1); 3) Theorem 5.3 is Theorem 2.4 in [35]. Their proofs are in §§ 2, 3–4 and 5 respectively. The method used in these proofs will be, in short, to modify the self-mapping f step by step so that in each step there is a homotopy between the initial and terminal self-mappings, and that the homotopy constructed is often one of the two types named in the title of § 1. The two Examples 3.4 and 3.5 are also of interest.

Let us note that in the present chapter simplex always means an *open simplex*. In order to avoid confusion we shall denote an open simplex by σ or s, to distinguish, $\bar{\sigma}$ or $\bar{s}$ for a closed simplex as used in Appendix C and [8]. Let us note also that we shall regard our complex K as lying in a Euclidian space of proper dimension, and thus we can talk of spherical neighborhoods and line segments in $|K|$. As usual, use $[a, b]$ for a closed segment. The broken line or broken line path formed by $[a, b]$ and $[b, c]$ is denoted by $[a, b, c]$. As the parameter t of the moving point p on a broken line path $[a, b, c, \cdots, d]$ we take the quotient of the length along the path from a to p by the total length of the path. Thus an orientation of the path is specified by the increasing parameter. As usual $U(a, \varepsilon)$ denotes the spherical neighborhood with center a and radius ε, and $\overline{U}(a, \varepsilon)$ its closure in $|K|$, while $U([a, b], \varepsilon) = \bigcup_{x \in [a,b]} U(x, \varepsilon)$. Reading Appendix C is a prerequisite for the present chapter.

1. Point-tail homotopy and line-fence homotopy

The present section is devoted to the definitions of the two types of homotopies named in the section title. They are the technical tools used to modify the self-mapping f of our polyhedron $|K|$. The discussion of the change of fixed point sets under these homotopies will be postponed to the succeeding sections.

1.1 *Lemma. Let K be a connected finite simplicial complex, $f\colon |K| \to |K|$ a self-mapping of $|K|$, point $a \in |K|$, and $p\colon I \to |K|$ a path on $|K|$ with $f(a) = p(0)$ as the initial point. Then, for positive ε not greater than some $\delta > 0$, there exists a homotopy $f_t\colon |K| \to |K|$ with the following properties:*

(i) $f_t\colon f_0 = f \simeq f_1$ rel $|K| - U(a, \varepsilon)$[1];

(ii) $f_1(x) = f((\frac{2}{\varepsilon}d(a,x) - 1)x + (2 - \frac{2}{\varepsilon}d(a,x))a)$, *when* $\frac{\varepsilon}{2} \leqslant d(a,x) \leqslant \varepsilon$;

(iii) $f_1(x) = p(1 - \frac{2}{\varepsilon}d(a,x))$, *when* $x \in \overline{U}(a, \frac{\varepsilon}{2})$;

(iv) *if* $\mathrm{Car}_k\, p(t) \cap \mathrm{Car}_k\, a \neq \emptyset$, $\forall t \in I$, *then* f_1 *possesses the property* S *(Definition* C 1.3) *on* $\overline{U}(a, \varepsilon)$.

Proof. There exists a sufficiently small number $\delta > 0$ such that $U(a, \delta) \subseteq \mathrm{st\ car}\ a$ (open star of the open carrier of the point a in K). When the additional hypothesis in (iv) holds, the hypothesis $f(a) = p(0)$ means that f possesses the property S at the point a. From the remark just before Definition C 1.3, we may take δ so small such that f possesses the property S on $\overline{U}(a, \delta)$. Hereafter we assume that δ meets these two requirements.

Take a fixed $\varepsilon > 0$, but $\leqslant \delta$. For any given $0 < t < 1$, $\overline{U}(a, \varepsilon t)$ is divided into three regions by the two "circles" $d(a,x) = \varepsilon t$ and $d(a,$

---

1) This expression means $f_t\colon f_0 \simeq f_1$ on $|K|$, and $f_t(x) = f_0(x), \forall x \in |K| - U(a, \varepsilon), t \in I$.

$x)=\varepsilon t/2$ (Figure 2, left). In the three corresponding regions of $|K|\times I$ (Figure 1), we define $f_t(x)$ as follows:

$$f_t(x)=\begin{cases} f(x), & \text{when } x\in|K|-U(a,\varepsilon t);\\ f\left(\left(\frac{2}{\varepsilon t}\,d(a,x)-1\right)x+\left(2-\frac{2}{\varepsilon t}\,d(a,x)\right)a\right), & \\ & \text{when } 0<\frac{\varepsilon t}{2}\leqslant d(a,x)\leqslant \varepsilon t;\\ p\left(t-\frac{2}{\varepsilon}d(a,x)\right), & \text{when } 0\leqslant d(a,x)\leqslant\frac{\varepsilon t}{2}. \end{cases} \tag{1}$$

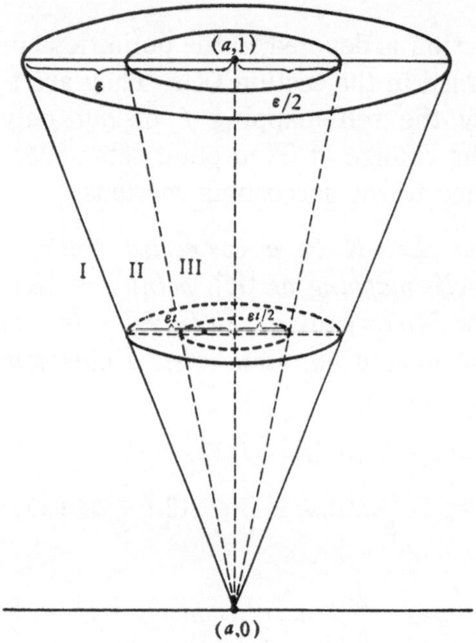

Figure 1

In the interior of each region, $f_t(x)$ is obviously continuous. When the point $(a,0)$ is left out, at any point of the common boundary

$$\{(x,t)\colon\ d(a,x)=\varepsilon t,\ t>0\}$$

of the regions I and II, the first and the second expressions in (1) take the same value, and at any point of the common boundary

$$\{(x,t)\colon\ d(a,x)=\frac{\varepsilon t}{2},\ t>0\}$$

of the regions II and III the second and the third expressions in (1) take the same value. The point $(a,0)$ is the only common point of the boundaries of the three regions. When (x,t) approaches $(a,\ 0)$ in $|K|\times I$, it is easy to see that $f_t(x)$ has the same limit $f(a)$. Thus from the continuity Lemma ([3] p. 62) $f_t(x)$ is continuous in $|K|\times I$ and is a homotopy as desired.

Remark 1. Before continuing and completing the proof, let us make one observation about the formula (1). When we set, for $t>0$,

$$\lambda(x,t)=\frac{2}{\varepsilon t}d(a,x)-1, \tag{1a}$$

$$x'=x'(x,t)=(1-\lambda(x,t))a+\lambda(x,t)x,$$

then the function on the right hand side of the second expression in (1) is simply $f(x'(x,t))$, where $x'(x,t)$ is the point dividing the closed $[a, x]$ in the ratio $\lambda(x,t)$:

$$(x'-a):(x-a)=\lambda(x,t),\ 0\leqslant\lambda(x,t)\leqslant 1.$$

Thus x' is a point of $[a,x]$.

When $t=1$, we find

$$f_1(x)=\begin{cases} f(x), & \text{when } x\in|K|-U(a,\varepsilon);\\ f(x'(x,1)), & \text{when } \frac{\varepsilon}{2}\leqslant d(a,x)\leqslant\varepsilon;\\ p(1-\frac{2}{\varepsilon}d(a,x)), & \text{when } 0\leqslant d(a,x)\leqslant\frac{\varepsilon}{2}; \end{cases} \tag{2}$$

where $x'(x,1)$ is given in (1a). The properties (i) to (iii) of f_1 are rather obvious. We only need to prove (iv).

Because the number δ chosen meets the second requirement, we have $\mathrm{Car}\,f(x)\cap\mathrm{Car}\,x\neq\emptyset$, $\forall x\in\overline{U}(a,\delta)$ and in particular, $\mathrm{Car}\,f(x')\cap\mathrm{Car}\,x'\neq\emptyset$, since $x'\in[a,x]\subseteq\overline{U}(a,\delta)$ as shown in Remark 1.

Consider first the case $\varepsilon/2\leqslant d(a,x)\leqslant\varepsilon$. For this case, from above $\mathrm{Car}f(x')\cap\mathrm{Car}x'\neq\emptyset$; $x'\in(a,x)\Rightarrow\mathrm{Car}x'=\mathrm{Car}x$; and the second expression in (2) gives $f_1(x)=f(x'(x,1))$. These three facts $\Rightarrow$

$$\mathrm{Car}f_1(x)\cap\mathrm{Car}\,x\neq\emptyset,\ \text{when}\ \frac{\varepsilon}{2}\leqslant d(a,x)\leqslant\varepsilon.$$

Next, consider the case $0\leqslant d(a,x)\leqslant\varepsilon/2$. Because the δ chosen meets the first requirement, $\mathrm{Car}a$ is a face of $\mathrm{Car}x$ in this case. This fact, together with the hypothesis in (iv) and the third expression in (2)$\Rightarrow f_1$ possesses the property S, when $0\leqslant d(a,x)\leqslant\varepsilon/2$. □

Remark 2. A special feature of the homotopy f_t: $f\simeq f_1$ is that the region $\overline{U}(a,\varepsilon t)$ appearing in the definition of the mapping f_t, $t\in I$, varies with t, while $f_0=f$. Figure 2 shows pictorially this homotopy. The two dotted and the two undotted circles represent $\partial U(a,\varepsilon t)$, $\partial U(a,\varepsilon t/2)$, for $0<t<1$, and $\partial U(a,\varepsilon)$, $\partial U(a,\varepsilon/2)$ respectively; the dotted and undotted contours on the right stand for the f-images of the corresponding circles. The four circles in Figure 2 may be regarded as the vertical projections of the two horizontal sections in Figure 1.

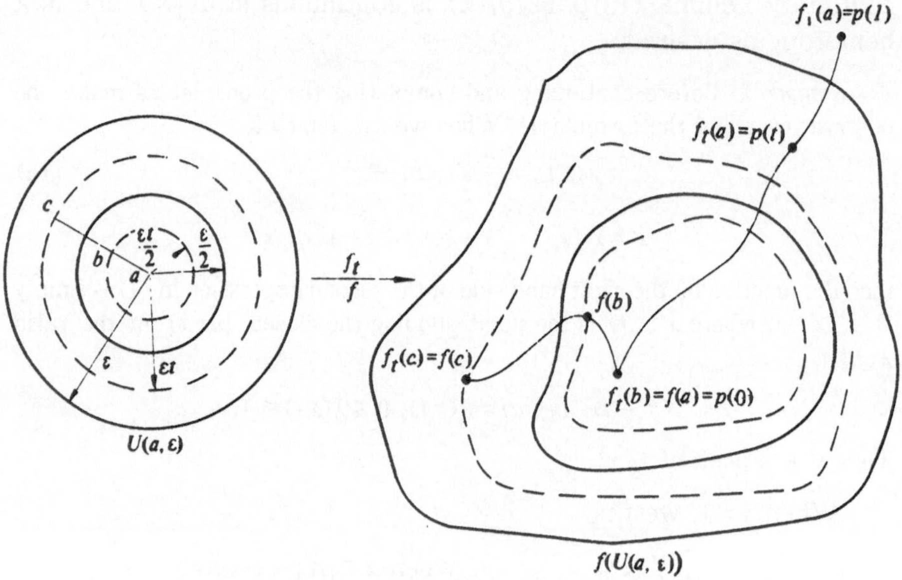

Figure 2 f and f_t

As $d(a,x)$ plays an important role in the second and the third expressions for f_t, consider the dotted circles and a running radius $[c,b,a]$ *of* $\overline{U}(a,\varepsilon t)$, where the point b is the middle point of $[c,a]$ and divides $[c,a]$ into the first half $[c,b]$ and the second half $[b,a]$. The f_t on $[c,b,a]$ (the third and the second expressions in (1)) may be described as follows. When x describes the oriented second half $[b,a]$, $d(a,x)$ decreases from $\varepsilon t/2$ to 0, and when $f_t(x)$ describes p_0^t (see Definition A 2.3) $d(a,x)$ decreases from $p(0)=f(a)$ to $p(t)=f_t(a)$. When x describes the oriented first half $[c,b]$, $d(a,x)$ decreases from εt to $\varepsilon t/2$, $x'(x,t)$ describes $[c,a]$, and $f_t(x)=f(x'(x,t))$. This can be loosely but conveniently summarized in a single sentence: To obtain f_t from f on $[c,b,a]$, pull out $f([b,a])$ and glue it to the subpath p_0^t, while stretching $f([c,b])$ to cover $f([c,a])$.

For the particular case $t=1$, the undotted circles take the place of the dotted ones.

1.2 Definition. The homotopy f_t in Lemma 1.1 is called the *point- tail homotopy* of f. This name is otilized as it might bring out better the underlying geometrical meaning (shown in Figure 2): Obtain f_t from f by attaching the subpath p_0^t as a tail at the point $f(a)$ of $f(U(a,\varepsilon t))$, $t\in I$. When $t=1$, the tail is the path p and results in f_1.

What we have discussed so far regards the point–tail homotopy. Prior to the discussion of the second homotopy named in the section title, we introduce the following definition and notations.

1.3 Definition. Let σ be a maximal simplex of K of dimension > 1, and points $a \in \sigma$, $b \in \bar{\sigma}$ but b not a vertex of K. There exists obviously a number $\varepsilon > 0$ such that $U([a,b],\varepsilon) \subseteq \mathrm{st\ car}\ b$. A contraction

$$R:\ U([a,b],\varepsilon) \to [a,b],\ x \mapsto R(x)$$

is a mapping defined as follows:

(i) When $b \in \sigma$, then $R(x) \in [a,b]$ and $d(x,R(x))$ is minimal (Figure 4, left).

(iia) When $b \in \partial\sigma$ (Figure 3) and $x \in \sigma$, let the ray starting at the point x and running along the direction $\overrightarrow{ab}$ intersect $\partial\sigma$ in a point y. Take the point z on the straight line ab such that $\overrightarrow{bz} = \overrightarrow{yx}$. Then define $R(x) = z$ or a according as $z \in$ or $\notin [a,b]$ (Figure 3 and Figure 4, left).

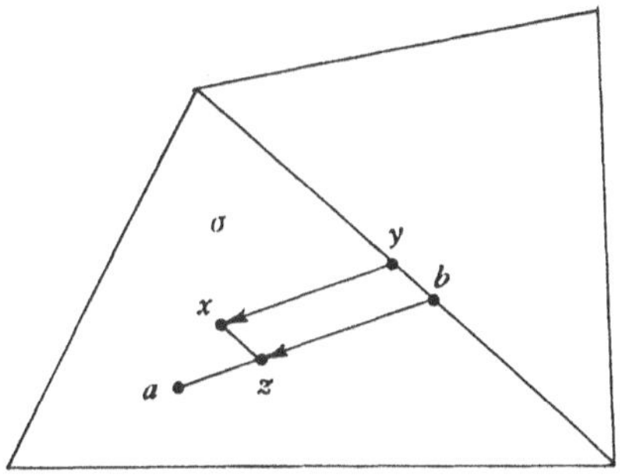

Figure 3

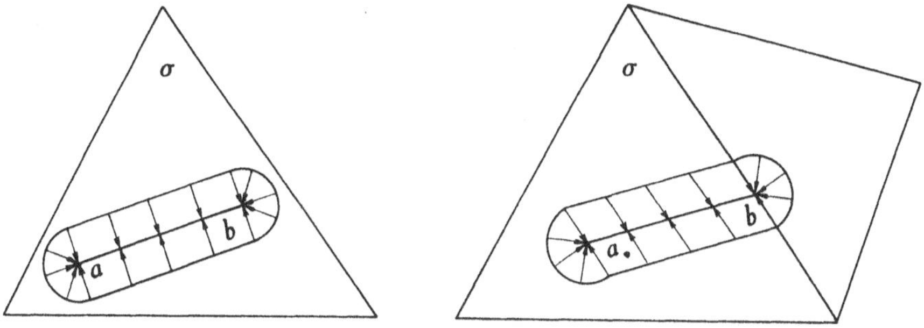

Figure 4 $W([a,b],\ \varepsilon)$

(iib) When $b \in \partial\sigma$ and $x \notin \sigma$, define $R(x) = b$ (Figure 4, right).

1.4 Notations. Let $R(x)$ be defined as in Definition 1.3 and M be a given subset of $[a,b]$. Define (see Figure 4)

$$W([a,b],\varepsilon) = \{x \in U([a,b],\varepsilon):\ d(x,\ R(x)) < \varepsilon\},$$

$$W([a,b],\varepsilon,\ M) = \{x \in W([a,b],\varepsilon):\ R(x) \in M\}.$$

It is easy to see that the set $W([a,b],\varepsilon)$ shown in Figure 4 left, is $U([a,b],\varepsilon)$, while that shown in Figure 4 right, is, in general, a proper subset of $U([a,b],\varepsilon)$.

1.5 Lemma. *Let K be a connected finite simplicial complex, σ a maximal simplex of K of dimension >1, and points $a\in\sigma$, $b\in\bar{\sigma}$ but b not a vertex of K. Again let $f\colon |K|\to|K|$ and $H\colon [a,b]\times I\to|K|$ be mappings such that $H(x,0)=f(x)$, $\forall x\in[a,b]$. Finally suppose M and N are subsets of $[a,b]$. Then, for positive ε not greater than some $\delta>0$, there exists a homotopy $f_t\colon |K|\to|K|$ with the following properties:*

(i) $f_t\colon f_0=f\simeq f_1$ rel $|K|-W([a,b],\varepsilon)$;

(ii) $f_1(x)=f\left(\left(\frac{2}{\varepsilon}d(x,R(x))-1\right)x+\left(2-\frac{2}{\varepsilon}d(x,R(x))\right)R(x)\right)$, *for* $\frac{\varepsilon}{2}\leqslant d(x,R(x))\leqslant\varepsilon$;

(iii) $f_1(x)=H\left(R(x),\ 1-\frac{2}{\varepsilon}d(x,R(x))\right)$, *for* $0\leqslant d(x,R(x))\leqslant\frac{\varepsilon}{2}$;

(iva) *If* $\operatorname{Car}H(x,\tau)\cap\operatorname{Car}x\neq\varnothing$, $\forall(x,\tau)\in M\times I$, *then f_1 possesses the property S on* $W([a,b],\varepsilon,M)$;

(ivb) *If* $d(H(x,\tau),x)>0$, $\forall(x,\tau)\in N\times I$, *then f_1 has no fixed point on* $W([a,b],\varepsilon,N)$.

Remark 3. This lemma is apparently a generalization of Lemma 1.1. Corresponding respectively to the point a, the neighborhood $U(a,\varepsilon)$, the path p and the property (iv) in Lemma 1.1 we have here the line segment $[a,b]$, the neighborhood $W([a,b],\varepsilon)$, the mapping H and the refined properties (iva) and (ivb). But the radius $[c,b,a]$ of the dotted circles in Figure 2 is replaced by $[x_2,x_1,x_0]$, $x_0=R(x_2)$, in the simplified Figure 5. Just as how to obtain f_t from f in

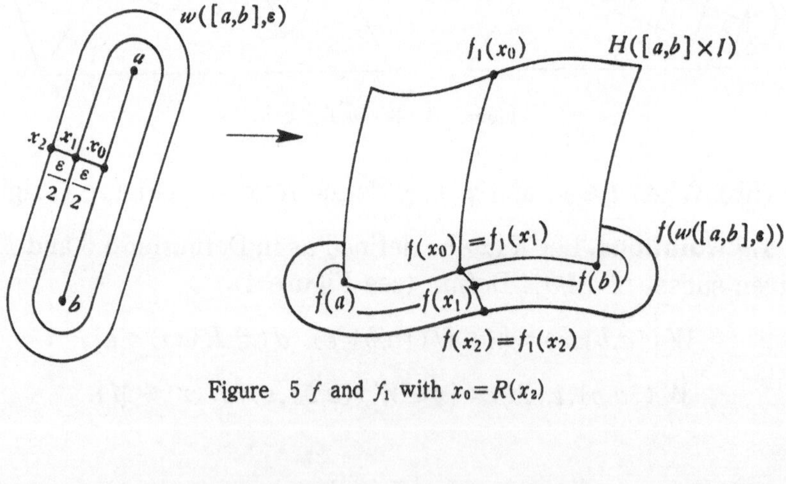

Figure 5 f and f_1 with $x_0=R(x_2)$

Lemma 1.1 was summarized in a single sentence in Remark 2, how to obtain f_1 from f in the present Lemma can be done similarly. We omit the details.

Proof. There exists a sufficiently small number $\delta>0$ such that 1) $U([a,b],\delta)\subseteq$ st car b. When the additional hypothesis in (iva) holds, the hypothesis $H(x,0)=f(x)$, $\forall x\in[a,b]$, means that f possesses the property S on M. From the remark just before Definition C 1.3, we can take δ so small such that f possesses the property S on $W([a,b],\delta,M)$.

When the additional hypothesis in (ivb) holds, then in particular for $\tau=0$ we have

$$d(f(x),x)>0,\ \forall x\in N.$$

Thus under this additional hypothesis, we can also take δ so small that

$$d(H(x,\tau),x)>\delta, \forall x\in N,\ 0\leqslant\tau\leqslant 1, \tag{3}$$

$$d(f(x),x)>\delta,\ \forall x\in W([a,b],\delta,N). \tag{4}$$

Set for $t>0$ (cf. (la))

$$\begin{aligned}\lambda(x,t)&=\frac{2}{\varepsilon t}d(R(x),x)-1,\\ x'(x,t)&=(1-\lambda(x,t))\ R(x)+\lambda(x,t)x.\end{aligned} \tag{4 a}$$

Then define f_t: $|K|\to|K|$ as follows:

$$f_t(x)=\begin{cases} f(x), \text{when } x\in|K|-W([a,b],\varepsilon t);\\ f(x'(x,t)), \text{when } 0<\dfrac{\varepsilon t}{2}\leqslant d(x,R(x))\leqslant\varepsilon t;\\ H\left(R(x),\ t-\dfrac{2}{\varepsilon}d(x,R(x))\right), \text{when } 0\leqslant d(x,R(x))\leqslant\dfrac{\varepsilon t}{2}.\end{cases} \tag{5}$$

The proof of continuity of $f_t(x)$ in the two variables (x,t) is similar to that in the proof of Lemma 1.1. Note that the common boundary of the three regions now is no longer a single point, but is $\{(x,0):x\in[a,b]\}$. There follows then

$$f_1(x)=\begin{cases} f(x), \text{when } x\in\ |K|-W([a,b],\varepsilon);\\ f(x'(x,1)), \text{when} \dfrac{\varepsilon}{2}\leqslant d(x,R(x))\leqslant\varepsilon;\\ H\left(R(x),\ 1-\dfrac{2}{\varepsilon}d(x,R(x))\right), \text{when } 0\leqslant d(R(x),x)\leqslant\dfrac{\varepsilon}{2}.\end{cases} \tag{6}$$

The properties (i) to (iii) of f_1 are also obvious. The proof of property (iva) is parallel to that of (iv) in Lemma 1.1.

It remains to prove only the property (ivb) of f_1. Let $x \in W([a, b], \varepsilon, N)$. For $\frac{\varepsilon}{2} \leqslant d(R(x),x) \leqslant \varepsilon$, from (6), (4) and (4a) it follows

$$\begin{aligned} d(f_1(x),x) &= d(f(x'(x,1)),\ x) \\ &\geqslant d(f(x'(x,1)), x'(x,1)) - d(x'(x,1),x) \\ &\geqslant \varepsilon - d(x'(x,1),x) > 0. \end{aligned}$$

For $0 \leqslant d(R(x),x) \leqslant \varepsilon/2$, set

$$\tau_0 = 1 - \frac{2}{\varepsilon} d(R(x),x);$$

then from (6) and (3) it follows

$$\begin{aligned} d(f_1(x),x) = d(H(R(x),\tau_0),x) &\geqslant d(H(R(x),\tau_0),\ R(x)) \\ &- d(R(x),x) \geqslant \varepsilon - d(R(x),x) > 0. \end{aligned}$$

Hence f_1 has no fixed point on $W([a,b],\varepsilon,N)$. □

1.6 Definition. The homotopy f_t in Lemma 1.5 is called the *line-fence homotopy* of f. Geometrically speaking, we obtain f_t from f by attaching $H:[a,b] \times [0,\varepsilon t] \to |K|$, $t \in I$, as a fence along the line $f([a,b])$.

2. Moving and uniting of fixed points. $\#\Phi(\langle id \rangle)$ of 2-dimensionally connected polyhedron

In the present section, we shall discuss the moving and uniting of the fixed points of a self-mapping f in a region, on which f possesses the property S. Lemma 2.1 will be used to unite the fixed points near a given point, or to move the fixed points in a maximal open simplex, and Lemma 2.2 to move a fixed point from one maximal open simplex to a neighboring one. By means of these two lemmas we shall determine $\#\Phi(\langle id \rangle)$ of 2-dimensionally connected polyhedron (Definition 2.3).

2.1 Lemma. *Let K be a finite connected simplicial complex, point $a \in |K|$, $\overline{U}(a,\varepsilon) \subseteq \text{stcar}\,a$. Let the self-mapping $f\colon |K| \to |K|$ have no fixed point on $\partial U(a,\varepsilon)$, and possess the property S on $\overline{U}(a,\varepsilon)$. Then there exists a self-mapping $g\colon |K| \to |K|$ with the following properties:*

(i) $g \simeq f$ rel $|K| - U(a,\varepsilon)$;
(ii) *g has only one fixed point at a in $U(a,\varepsilon)$;*
(iii) *g possesses the property S on $\overline{U}(a,\varepsilon)$.*

Proof. Since f has no fixed point on $\partial U(a,\varepsilon)$, there exists a number η, $0<\eta<\varepsilon$, such that f has no fixed point on $\overline{U}(a,\varepsilon)-U(a,\eta)$. Since f possesses the property S on $\overline{U}(a,\varepsilon)$, we have the function $\alpha(x,f(x),t)$, $\forall (x,t)\in \overline{U}(a,\varepsilon)\times I$, explained in Theorem C 1.1. Let $D(t)$ be a function on $[0,\varepsilon]$ which increases monotonically from $D(0)=0$ to $D(\varepsilon)=1$ and whose values are sufficiently small for $t\in[0,\eta]$ such that

$$\alpha(x,f(x),\ D(d(x,a)))\in \operatorname{st car} a,\ \forall\ x\in \overline{U}(a,\eta). \tag{1}$$

Take the following self-mapping f_1: $|K|\to|K|$ (see Figure 6 and Theorem C 1.1):

$$f_1(x)=\begin{cases} f(x), & \text{when } x\in|K|-U(a,\varepsilon);\\ \alpha(x,f(x),\ D(d(x,a))), & \text{when } x\in U(a,\varepsilon)-a;\\ a, & \text{when } x=a.\end{cases} \tag{2}$$

For any point $x\in\overline{U}(a,\eta)-a$, the ray starting at a and passing through x intersects $\partial U(a,\eta)$ in a unique point $v(x)$, and there is a unique value $t(x)$, $0<t(x)\leqslant 1$ such that

$$x=(1-t(x))a+t(x)v(x).$$

Both $v(x)$ and $t(x)$ are continuous in $x\in\overline{U}(a,\eta)-a$. Finally, define

$$g(x)=\begin{cases} f_1(x), & \text{when } x\in|K|-U(a,\eta);\\ (1-t(x))a+t(x)f_1(v(x)), & \text{when } x\in\overline{U}(a,\eta)-a;\\ a, & \text{when } x=a.\end{cases} \tag{3}$$

From (1), one checks $f_1(v(x))\in$ st car a, and hence in (3) we have

$$(1-t(x))a+t(x)\ f_1(v(x))\in \text{st car } a. \tag{4}$$

(i) Since f and f_1 possess the property S (see Definition C 1.2) on $|K|$ and $f(x)=f_1(x)$, $\forall\ x\in|K|-U(a,\varepsilon)$, from Theorem C 1.5 there follows $f_1\simeq f$ rel $|K|-U(a,\varepsilon)$. Similarly, f_1 and g possess the property S on $|K|$, and $f_1(x)=g(x)$, $\forall x\in|K|-U(a,\eta)$; and hence, from Theorem C 1.5 again, there exists $f_1\simeq g$ rel $|K|-U(a,\eta)$. Thus $f\simeq g$ rel $|K|-U(a,\varepsilon)$.

(ii) From (2) and Theorem C 1.1, f_1 has no fixed point in $\overline{U}(a,\varepsilon)-U(a,\eta)$; from (3), $f_1(x)=g(x)\neq x$, $\forall x\in\overline{U}(a,\varepsilon)-U(a,\eta)$. But from (3) over again, g has the only fixed point a in $U(a,\eta)$.

(iii) From (2) and Theorem C 1.1 (iv), f_1 possesses the property S on $\overline{U}(a,\varepsilon)$. Again from $\overline{U}(a,\eta)\subseteq$ st car a and (4), g possesses the property S on $\overline{U}(a,\varepsilon)$. □

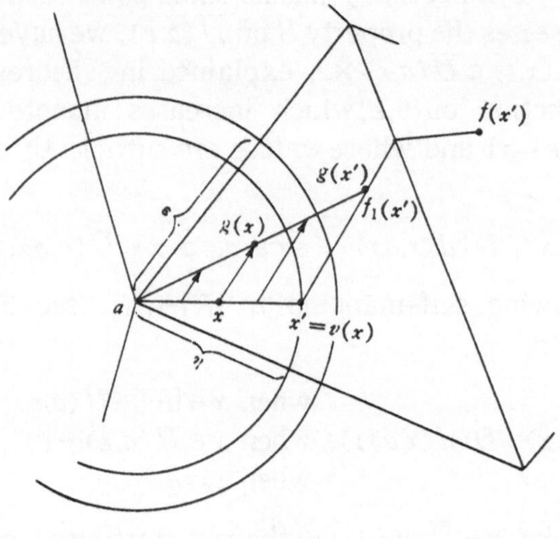

Figure 6

Remark 1. Let us note: in the statement of our lemma, we do not specify whether or not the point a is a fixed point of f, but it is a fixed point of both f_1 and g.

2.2 Lemma. *Let K be a connected finite simplicial complex, σ_1 and σ_2 its two maximal simplexes of dimensions be greater than* 1, *point $a \in \sigma_1$, point $b \in \bar{\sigma}_1 \cap \bar{\sigma}_2$ but not a vertex of K, and f: $|K| \to |K|$ a self-mapping of $|K|$ with the following properties*:

1) *a is an isolated fixed point of f, and f has no other fixed point in $[a,b]$;*

2) *f possesses the property S on $[a,b]$.*

Then, for a sufficiently small number $\varepsilon > 0$ and any given point $c \in U(b,\varepsilon) \cap \sigma_2$, there exists a mapping g: $|K| \to |K|$ with the following properties:

(i) $g \simeq f$ rel $|K| - U([a,b],\varepsilon)$;

(ii) *g has only one fixed point c in $U([a,b],\varepsilon)$;*

(iii) *g possesses the property S on $\overline{U}([a,b],\varepsilon)$.*

Proof. From the two properties of f, we can take a sufficiently small number $\varepsilon > 0$ such that $\overline{U}(b,\varepsilon) \subseteq \text{st car } b$; $f(x) \notin U(b,\varepsilon)$, $\forall x \in U(b,\varepsilon)$; f has no other fixed point in $\overline{U}([a,b],\varepsilon)$ besides a; and f possesses the property S on $\overline{U}([a,b],\varepsilon)$.

Take a broken-line path p from a point e in σ_2 to a point d on (a,b) such that $p(t) \in \text{st car } b$, $p(t) \notin \overline{U}(b,\varepsilon)$, $t \in I$ (Figure 7). Since f possesses the property S at point b and $f(b) \notin U(b,\varepsilon)$, we can take

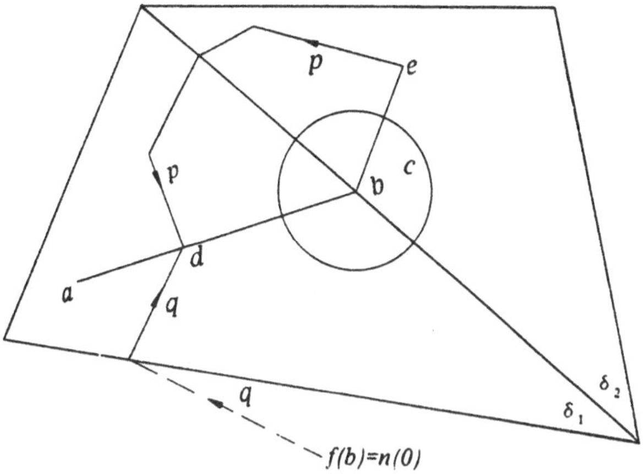

Figure 7

a broken-line path q from $f(b)$ to d such that q and $\overline{U}(b,\varepsilon)$ have no common point and

$$\text{Car } q(t) \cap \text{Car } b \neq \emptyset, t \in I. \tag{5}$$

Let us note that the hypothesis $b \neq a$, vertex of K ensures the existence of the paths p and q. Set

$$r = q([d,b,e]p)^j,$$

where e and d are arbitrary points of σ_2 and $[a,b]$ respectively, but $\notin \overline{U}(b,\varepsilon)$, and $j = v(f, a)$ is the index of the isolated fixed point a of f (Theorem II 5.1). Since $r(0) = f(b)$, on applying Lemma 1.1 to the mapping f and the path r, we obtain a mapping f_1: $|K| \to |K|$ such that

$$f_1 \simeq f \text{ rel } |K| - U(b,\varepsilon),$$

and by virtue of (5) and Lemma 1.1 (iv), f_1 possesses the property S on $\overline{U}(b,\varepsilon)$. In $U(b,\varepsilon) \cap \bar{\sigma}_1$, f_1 has only the fixed points $d_1, \cdots, d_{|j|}$, all lying in (d,b) (Figure 8). Since the f_1-image of a neighborhood of d_i lies in $[d,b]$, from Theorem II 5.1 (va) and Definition I 4.2 there follows $v(f_1, d_i) = -\text{sgn}(j)$, $i = 1, \cdots, |j|$. Thus

$$\sum_{i=1}^{|j|} v(f_1, d_i) = -|j| \text{ sgn}(j) = -j.$$

Similarly, the fixed points of f_1 on $U(b,\varepsilon) \cap \bar{\sigma}_2$ are $e_1, \cdots, e_{|j|}$, all in (b, ε), and

$$\sum_{i=1}^{|j|} v(f_1, e_i) = j.$$

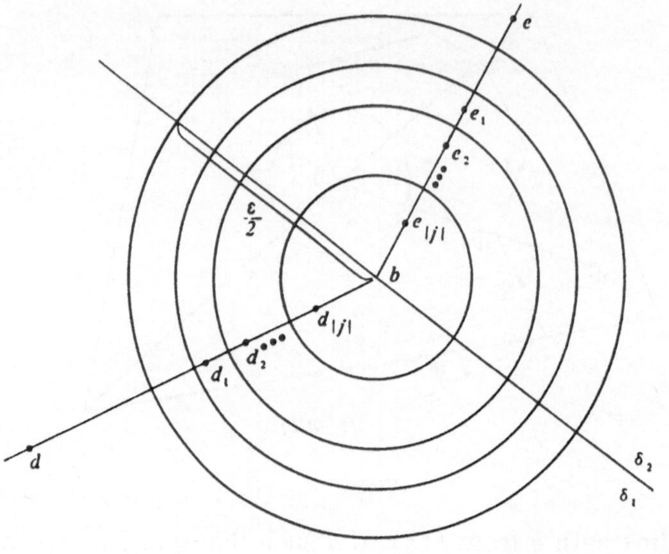

Figure 8

As f_1 possesses the property S on $\overline{U}(b,\varepsilon)$, and f does also on $\overline{U}([a,b],\varepsilon)$, so does f_1 also on $\overline{U}([a,b],\varepsilon)$. On applying Lemma 2.1 to σ_1, we have the fixed points $a, d_1,\cdots,d_{|j|}$ moved and united onto the single fixed point a with the index 0. On applying Theorem II 5.1 (vi) again, we have the fixed point a removed but the property S preserved. Finally, on applying Lemma 2.1 to moving and uniting the fixed points $e_1,\cdots,e_{|j|}$ onto the arbitrarily given point c of $\sigma_a \cap U(b,\varepsilon)$, we obtain the desired mapping g. □

Remark 2. In Figures 7 and 8, $\sigma_1 \cup \sigma_2$ is represented by a quadrangle and the locus $d(b,x)=a$ positive constant by a circle. Of course this is the case only when dim $\sigma_1=$dim $\sigma_2=2$. The circular region in Figure 7 represents $0 \leqslant d(b,x) \leqslant \varepsilon$, while the largest circular region in Figure 8 represents $0 \leqslant d(b,x) \leqslant \varepsilon/2$. These circles correspond to those in Figure 2 at the left, but r corresponds to p in Figure 2 at the right.

f_1 maps the region $\frac{\varepsilon}{2} \geqslant d(b,x) \geqslant 0$ onto r, the circle $d(b,x)=\frac{\varepsilon}{2}$ onto the point $f(b)=r(0)$, the circular ring bounded by this circle and the next smaller one onto the subpath q of r and the next smaller ring as well as its intersection segment with $[d,b]$ onto the first loop $([d,b,e]p)^{\mathrm{sgn}(j)}$ of r. Hence f_1 has a fixed point d_1 in this intersection segment, etc.

Figures 7 and 8 are only geometrical illustration of our proof which is valid in general when dim σ_1 and dim $\sigma_2>1$.

The two lemmas above will lead us immediately to the important theorem on the least number of fixed points of the identity mapping class (Theorem 2.4).

2.3 Definition. Let K be a connected finite simplicial complex. If each of its maximal simplexes is of dimension $\geqslant 2$, and every pair of its 2–simplexes σ' and σ'' can be joined by such a sequence of 2–simplexes $\sigma'=\sigma_1,\sigma_2,\cdots,\sigma_s=\sigma''$, that every two successive ones σ_i and σ_{i+1} of which have a common 1–face, then K is called a **2–dimensionally connected complex.**

2.4 Theorem. *If K is a 2–dimensionally connected complex, then the least number of fixed points of the identity mapping class $\#\Phi(\langle id\rangle)$ equals $N(id)$, i.e., it is 0 or 1 according as the Euler–Poincar'e characteristic $\chi(K)$ of K is 0 or $\neq 0$.*

Proof. By virtue of the approximation Theorem C 2.5, there exists a simplicial mapping $f\in\langle id\rangle$ for any $\varepsilon>0$ such that $d(f(x),x)<\varepsilon, \forall x\notin|K|$, and all the fixed points of f are isolated and are points interior to the maximal simplexes of K. We can take ε smaller than the δ in Theorem C 1.4 to ensure that f possesses the property S on $|K|$. Since K is 2–dimensionally connected, on applying Lemmas 2.1 and 2.2 we can move and unite all the fixed points of f, that is obtain finally a mapping g: $|K|\to|K|$ with a single fixed point, $g\in\langle id\rangle$. If $\chi(K)=0$, we can moreover have this fixed point removed by virtue of Theorem II 5.1(v) and (vi).

$$\text{From } L(id)=\chi(K) \text{ and Example II 6.1 (ii),}$$

there follows at once the statement of our theorem. □

2.5 Corollary. *If K is a 2–dimensionally connected complex and $\chi(K)=0$, then there exists a self–mapping $f\in\langle id\rangle$ of $|K|$, free from fixed point. Moreover, $N(id)=\#\Phi(\langle id\rangle)=0$.* □

3. Non–2–dimensionally–connected complex. Welding set. Good star motion

In the preceding section, $\#\Phi(\langle id\rangle)$ of a 2–dimensionally connected complex was determined. In the present section we are going to study $\#\Phi(\langle id\rangle)$ of a general complex.

3.1 Definition. A finite connected simplicial complex K, which is not 2–dimensionally connected and not reduced to a simple vertex, is called a *non–2–dimensionally–connected complex.* Such a K must consist of a certain number of closed maximal 1–simplexes and a certain number of maximal 2–dimensionally connected subcomplexes are denoted by $M_i(K)$, or simpler M_i, $i=1,2,\cdots,k$, and are called *branches* of K. Set

$$\dot{M}_i(K) \text{ or } \dot{M}_i=|M_i|\cap|\overline{K-M_i}|,\ \dot{M}(K) \text{ or } \dot{M}=\bigcup_{i=1}^{k}\dot{M}_i.$$

Call $\dot{M}$ *the welding set* of K and $\dot{M}_i$ the *specified subsets* of $\dot{M}$.

When $k=1$, the complex $K=M_1$ must be either a single closed 1-simplex or a 2-dimensionally connected complex, and $\dot{M}_1=\emptyset$ in either case. When $k>1$, each $\dot{M}_i$ is a nonempty set of vertices of K.

Remark. In the present and the next section, we study only non-2-dimensionally-connected complexes *with more than one branches.* This is a limitation on the complex K, but not on the polyhedron $|K|$. In fact, a closed 1-simplex $[a, b]$ has only one branch, while its first subdivision, the broken-line $[a,c,b]$ is a complex with $K=2$ and $\dot{M}_i=\dot{M}=\{c\}$ nonempty. It is easy to see that for a non-2-dimensionally-connected complex K, the following three properties are equivalent: with $k>1$ branches, with a nonempty welding set, and with at least two 1-simplexes.

Moreover, the implication of the welding set $\dot{M}$ of a complex K with $k(>1)$ branches is three-fold. First, the set is a nonempty subset of vertices of K. Second, it is the union of $k(>1)$ nonempty specified subsets $\dot{M}_i$, any two of which may have common points. Finally, every $\dot{M}_i$ is a *separating set* of K; i.e., though $|M_i|$ and $|M_i|-\dot{M}_i$ are connected, $|K|-\dot{M}_i$ is not, and $|K|-\dot{M}$ consists of $k(>1)$ connected components.

Example 3.1. Let K_1 be the 1-dimensional complex or graph shown in Figure 9a). From K_1, we construct an n-dimensional complex K_2 as follows. Along each edge of K_1, glue an n-dimensional closed simplex $(n>1)$ so that the intersection of the n-simplex with K_1 is just that edge and that any two such attached n-simplexes have no intersection outside $|K_1|$. $\dot{M}(K_1)=\dot{M}(K_2)$ consists of all vertices of K_1, and is the union of 11 specified subsets. These two examples and their $\#\Phi(\langle id\rangle)$ were studied in [36 III] p. 574.

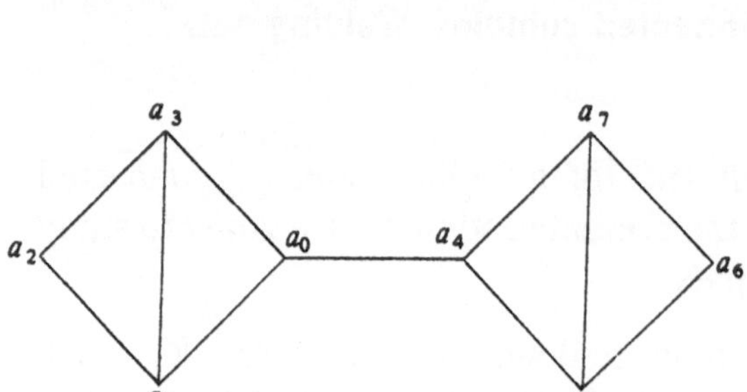

Figure 9 a) K_1, K_2

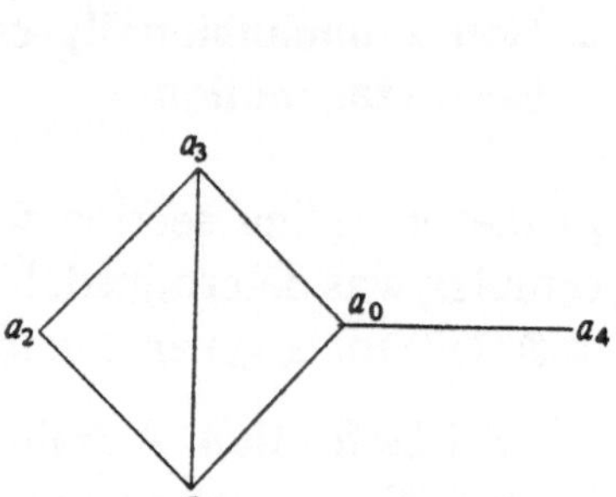

Figure 9 b) K_3

Example 3.2. The welding set $\dot{M}(K_3)$ of the graph K_3 shown in Figure 9b) is $\{a_0, a_1, a_2, a_3\}$, and is the union of 6 specified subsets, namely,

$$\{a_0, a_1\}, \{a_1, a_2\}, \{a_2, a_3\}, \{a_0, a_3\}, \{a_1, a_3\}, \{a_0\}.$$

3.2 Definition. Let K be a non-2-dimensionally-connected complex with more than one branches. To each $a \in \dot{M}(K)$, let $\tilde{g}(a)$ be either a itself or one definite branch M_i of K which contains a .Such a function $\tilde{g}$ is called a *star motion of the welding set* $\dot{M}(K)$. When $\tilde{g}(a)=a$, then a is called also a fixed point of $\tilde{g}$.

This concept of star motion $\tilde{g}$ of $\dot{M}(K)$ is an abstraction of the following more concrete concept of star mapping.

3.3 Definition. Let K be a non-2-dimensionally-connected complex with more than one branches. If g: $\dot{M}(K) \to |K|$ is a mapping such that 1) $g(a) \in \mathrm{st}_K a$, $\forall\ a \in \dot{M}(K)$ and 2) g: $\dot{M}(K) \to g(\dot{M}(K))$ is one to one, then g is called a *star mapping of the welding set.*

Such a star mapping g obviously gives rise to the following unique star motion of $\dot{M}(K)$, called the *induced star motion*:

$$\tilde{g}(a) = \begin{cases} a, & \text{when } g(a)=a; \\ M_i, & \text{when } g(a) \neq a \text{ and } g(a) \in |M_i|. \end{cases}$$

Conversely, given a star motion, one can easily construct a star mapping inducing it. The sets of fixed points of g and the induced $\tilde{g}$ are obviously the same: $\Phi(\tilde{g}) = \Phi(g)$.

3.4 Definition. Let $\tilde{g}$ be a star motion of the welding set $\dot{M}(K)$. $\tilde{g}$ is called a *good star motion* of $\dot{M}(K)$, if for each branch M_i, $i=1,2,\cdots,k$, either $\tilde{g}$ has at least one fixed point in $\dot{M}_i$, or $\tilde{g}$ assigns just $\chi(\mathrm{M}_i)$ of the points of the set $\dot{M}_i$ to branches $\neq M_i$.

Let us note that for a given branch M_i, there are the three following cases:

$$\# \dot{M}_i \geqslant \chi(M_i) \geqslant 0,\ \chi(M_i) > \# \dot{M}_i,\ \chi(M_i) < 0.$$

By our definition, a good star motion $\tilde{g}$ must have fixed point in each of the last two cases.

Example 3.3. In each of the complexes in Figure 10, let $g(a_j) = a_j'$ and $\tilde{g}$ be the star motion induced by the star mapping g. Figure 10a) shows a good star motion of $\dot{M}(K_1)$ as well as of $\dot{M}(K_2)$.

In Example 3.1, we explained how to construct K_2 from K_1 by gluing an n- simplex along each edge of K_1. Instead of an edge of an n-simplex, let us glue an edge of a 2-dimensionally connected complex M_i along the i-th edge of K_1. Moreover, take M_i such that each $\chi(M_i)$ is either $>2= \# \dot{M}_i$ or <0. Denote the conctructed complex by K_4. Then Figure 10a) shows a star motion but not a good star motion of $\dot{M}(K_4)$, while Figure 10c) shows a good star motion of $\dot{M}(K_4)$.

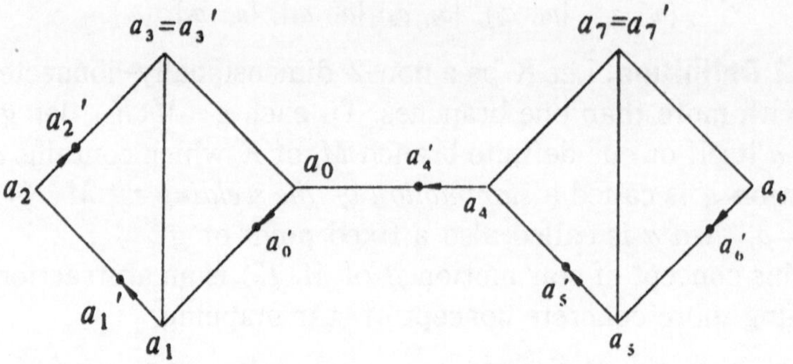

Figure 10 a) $\dot{M}(K_1)$, $\dot{M}(K_2)$; $\Phi(\tilde{g})=\{a_3, a_7\}$

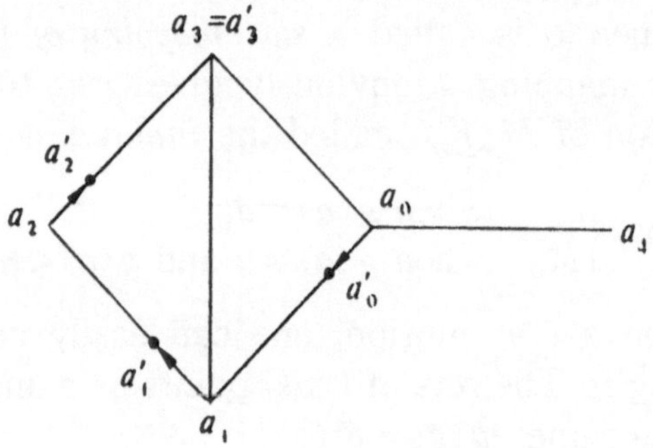

Figure 10 b) $\dot{M}(K_3)$; $\Phi(\tilde{g})=\{a_3\}$

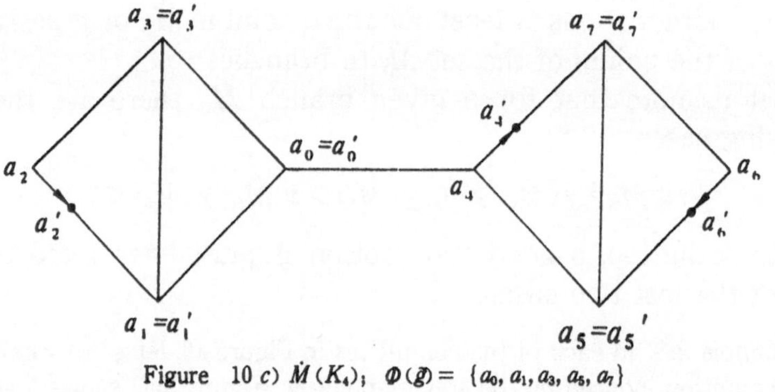

Figure 10 c) $\dot{M}(K_4)$; $\Phi(\tilde{g})=\{a_0, a_1, a_3, a_5, a_7\}$

A trivial example of a good star motion $\tilde{g}$ of a non-2-dimensionally-connected complex K with $k(>1)$ branches is one induced by the identity mapping $g=id$: $\dot{M}(K)\to|K|$. If a star motion $\tilde{g}$ has at least one fixed point on each $\dot{M}_i$, then $\tilde{g}$ is obviously a good star motion. As usual, let $\#\ \Phi(\tilde{g})$ be the number of fixed points of $\tilde{g}$. When $\tilde{g}$ runs through the set of all good star motions of the welding set $\dot{M}(K)$ of K, the greatest lower bound of $\#\Phi(\tilde{g})$ is called *the least number of fixed points of good star motions of the welding set* $\dot{M}(K)$ *of* K, and is denoted by $\# \varphi(K)$. The aim of the present and the next sections is to prove the following important theorem ([35a]).

3.5 Theorem (Shi). *If* K *is a non-2-dimensionally-connected complex with more than one branches, then*

$$\# \Phi(\langle id \rangle) = \# \varphi(K). \qquad \square$$

Example 3.4. It is easy to verify that the good star motion of $\dot{M}(K_1)$ or $\dot{M}(K_2)$ shown in Figure 10a) and those of $\dot{M}(K_3)$ and $\dot{M}(K_4)$ shown respectively in Figure 10b) and 10c) are good star motions with the least number of fixed points, i.e.

$$\# \varphi(K_1) = \# \varphi(K_2) = 2, \ \# \varphi(K_3) = 1, \ \# \varphi(K_4) = 5.$$

From Theorem 3.5, these numbers are also the correspending $\# \Phi(\langle id \rangle)$.

From Example II 6.1 (ii) and Theorem II 5.1 (iv), it follows that, for K_1 and K_2, $\# \Phi(\langle id \rangle) > N(id) = 1$. We leave it to the reader to decide whether these two numbers equal for K_3; and for K_4.

When K is a non-2-dimensionally-connected complex with more than one branches, its welding set $\dot{M}(K)$ is finite and nonempty. Thus the significance of the Shi Theorem 3.5 is that it renders the determination of $\# \Phi(\langle id \rangle)$ to the determination of $\# \varphi(K)$, a combinatorial problem.

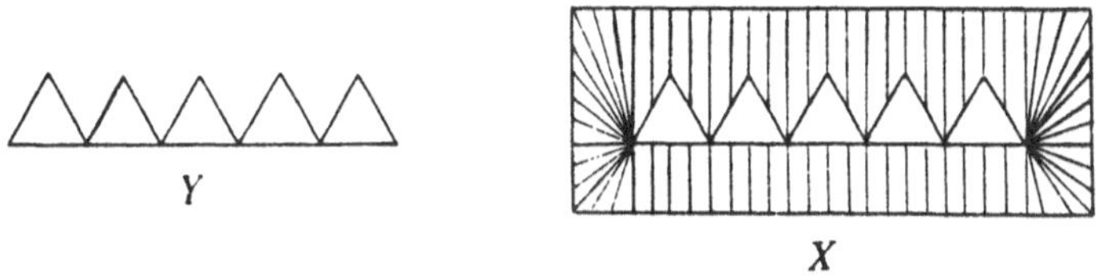

Example 3.5. Let K_1 be the graph shown on the left of the accompanying figure. It consists of a series of five triangles and is a non-2-dimensionally-connected complex with $k=15$ branches. Write $Y = |K_1|$. Shown on the right of the figure is a rectangular strip with a congruent K_1 in its interior and with the interiors of the five triangles removed. Let X be the polyhedron so obtained. Obviously X is a polyhedron of a 2-dimensionally connected complex. From Theorems 2.4 and 3.5, the least numbers of fixed points of the identity mapping classes for X and Y are respectively

$$\# \Phi(\langle id_X \rangle) = 1, \ \# \Phi(\langle id_Y \rangle) = 2.$$

On the other hand, let $j: Y \to X$ be the inclusion of Y into X and $r: X \to Y$ the retraction of X onto Y. Then $j \circ r \simeq id_X$: $X \to X$, $r \circ j = id_Y$: $Y \to Y$, and X and

Y are of the same homotopy type with (j,r) as a pair of homotopy equivalences. The identity mapping class $\langle id_X\rangle$ of X corresponds to the $\langle id_Y\rangle$ of Y under the correspondence induced by the pair (j,r) (Definition II 7.7). Thus the inequality of $\#\Phi(\langle id_X\rangle)$ and $\#\Phi(\langle id_Y\rangle)$ shows that *the least number of fixed points* $\#\Phi(\langle f\rangle)$ *of a mapping class* $\langle f\rangle$ *is not a homotopy type invariant.*

The following Lemmas 3.6a and 3.6 lead us to the key concept of a good star motion introduced in Definition 3.4.

3.6a Lemma. *Let K be a non-2-dimensionally-connected complex with more than one branches, and M_i one of the branches. If a mapping $f\colon |M_i|\to|K|$ possesses the property S on $|M_i|$, then the following*

$$g(x)=\begin{cases} f(x), & \text{when } f(x)\in|M_i|;\\ \mathrm{Car}_K x\cap \mathrm{Car}_K f(x), & \text{when } f(x)\notin|M_i|-\dot{M}_i \end{cases} \tag{1}$$

defines a mapping $g\colon |M_i|\to|M_i|$.

Proof. Our lemma aims at the construction of $g\colon |M_i|\to|M_i|$ from the given $f\colon |M_i|\to|K|$, which may be regarded as a generalization of the construction in [1], p. 534. We will prove that g is indeed a mapping.

First of all, the property S of f implies that the intersection $\mathrm{Car}_K x \cap \mathrm{Car}_K f(x)$ in the second expression of (1) is not empty and is a closed simplex of dimension $\geqslant 0$ of K, the common face of $\mathrm{Car}_K x$ and $\mathrm{Car}_K f(x)$. Now $x\in|M_i| \Rightarrow \mathrm{Car}_K x\in M_i$, and $f(x)\notin|M_i|-\dot{M}_i \Rightarrow \mathrm{Car}_K f(x)\in K-(M_i-\dot{M}_i)$. Thus the intersection $\in M_i\cap(K-(M_i-\dot{M}_i))=\dot{M}_i$ and must be a single welding vertex in $\dot{M}_i$. This shows that g is well defined.

g is defined separately on the following two closed subsets of $|M_i|$ (remember that $\dot{M}_i$ is a separating set of $|K|$):

$$R_1=f^{-1}(|M_i|),\ R_2=f^{-1}(|K|-(M_i-\dot{M}_i)),$$

$$R_1\cup R_2=|M_i|,\ R_1\cap R_2=f^{-1}(\dot{M}_i).$$

g is obviously continuous in R_1. Let us show the continuity of g in R_2 as follows. Set for $x\in R_2$

$$a=g(x)=\mathrm{Car}_K x\cap\mathrm{Car}_K f(x)\in\dot{M}_i.$$

There follow $x\in \mathrm{st}a$ and $f(x)\in \mathrm{st}a$. Since $\mathrm{st}a$ is an open subset of $|K|$ and f is continuous in $|K|$, there exists a neighborhood V of x in $|K|$ such that both V and $f(V)\subseteq \mathrm{st}a$. Then $U=V\cap R_2$ is an open set of R_2 and thus a neighborhood of x in R_2. Hence $f(y)\notin|M_i|-\dot{M}_i$, $\forall\, y\in U$; thus $g(U)=a$; in other words, g is continuous at x in R_2.

Finally, from the Continuity Lemma ([3] p.62), g is continuous in $|M_i|$. □

3.6 Lemma. *Let f: $|M_i| \to |K|$ and g: $|M_i| \to |M_i|$ be verbatim as those in Lemma 3.6a. If f has no fixed point on $\dot{M}_i$, and maps exactly ω_i vertices in $\dot{M}_i$ to points of $|K| - |M_i|$, then*

$$v(f, |M_i| - \dot{M}_i) = \chi(M_i) - \omega_i.$$

Proof. From (1), f and g on $|M_i| - \dot{M}_i$ have the same fixed point sets, and take on the same value at each point in the neighborhood $W = (f^{-1}(|M_i| - \dot{M}_i)) \cap (|M_i| - \dot{M}_i)$ of the fixed point sets. From Theorem II 5.1 (iii), there follows

$$v(f, |M_i| - \dot{M}_i) = v(f, W) = v(g, W) = v(g, |M_i| - \dot{M}_i). \tag{2}$$

From (1), the fixed points of g on $\dot{M}_i$ are just those ω_i vertices, say

$$\{a_j\},\ j = 1.2, \cdots, \omega_i$$

in $\dot{M}_i$, which are mapped by f to points of $|K| - |M_i|$. Moreover, from the argument at the end of the proof of Lemma 3.6a we see g maps a neighborhood $U(a_j, \delta_j)$ of a_j in $|M_i|$ onto a_j. From Theorem II5.1 (iii), we have

$$\begin{aligned} v(g, |M_i| - \dot{M}_i) &= v(g, |M_i| - \bigcup_{j=1}^{\omega_i} \overline{U}(a_j, \delta_{j/2})) \\ &= v(g, |M_i|) - \sum_{j=1}^{\omega_i} v(g, U(a_j, \delta_j)), \end{aligned} \tag{3}$$

and from Theorem II5.1 (iv), we have

$$v(g, U(a_j, \delta_j)) = 1,\ j = 1, 2, \cdots, \omega_i. \tag{4}$$

Finally from (1), g possesses the property S on $|M_i|$. Hence from Corollary C1.6, $g \simeq id$: $|M_i| \to |M_i|$, and

$$v(g, |M_i|) = \chi(M_i). \tag{5}$$

The conclusion of our lemma results from equations (2) to (5). □

3.7 Theorem. *If $\tilde{g}$ is a good star motion of the welding set $\dot{M}(K)$ of a non-2-dimensionally-connected complex K with more than one branches, then there exists a self-mapping G of $|K|$, which possesses the property S on $|K|$ and $\Phi(G) = \Phi(\tilde{g})$, and hence*

$$\# \Phi(\langle id \rangle) \leqslant \# \varphi(K).$$

Proof. First, let g: $\dot{M} \to |K|$ be a star mapping inducing the given $\tilde{g}$. Secondly, by virtue of Theorem C2.5, for any given number $\varepsilon > 0$, there exists a self-mapping $F \simeq id$ of $|K|$ such that $d(F(x),x) < \varepsilon$, $\forall\ x \in |K|$, and every fixed point of F is isolated and lies in a maximal simplex of a definite subdivision of K and hence also in a maximal simplex of K.

From Theorem C1.4, ε can be taken so small that F possesses the property S on $|K|$, and hence

$$F(a_j) \in \text{st}\ a_j,\ \forall a_j \in \dot{M}(K).$$

Take on $|K|$ the broken-line $[F(a_j),\ b_j,\ a_j,\ c_j,\ g(a_j)]$ as the path p_j, where b_j and c_j are so chosen that $\text{Car}_K b_j$ is a maximal simplex with $\text{Car}_K\ F(a_j)$ as a face, and that $\text{Car}_K c_j$ is a maximal simplex with $\text{Car}_K\ g(a_j)$ as a face.

For every $a_j \in \dot{M}(K)$, on applying Lemma 1.1 to attaching a tail p_j at $F(a_j)$ (see Definition 1.2) we obtain a mapping F_1: $|K| \to |K|$ such that F_1 possesses the property S on $|K|$, and its new fixed points are all isolated and lie on the segments $(b_j,\ a_j)$ and $(a_j,\ c_j)$. Thus the fixed points of F_1 in $|K| - \dot{M}(K)$ are all isolated and lie in maximal simplexes of K, and $F_1(a_j) = g(a_j)$, $\forall a_j \in \dot{M}(K)$.

Consider the following possible cases of the branch M_i with $\dot{M} = \dot{M}_i(K) = \{a_1,\ a_2, \cdots, a_h\}, h \geqslant 1$.

1) M_i is a closed 1–simplex, $h=1$ or 2. No matter whether g has none, one or two fixed points in $\dot{M}_i$, F_1 can be modified in the identity mapping class to possess the property S on $|M_i|$, to have no fixed point on $|M_i| - \dot{M}_i$, and to preserve the image $F_1(\dot{M}_i)$.

2) M_i is a 2–dimensionally connected complex, and g has no fixed point in $\dot{M}_i$. From Lemma 3.6 and the hypothesis that $\tilde{g}$ is a good star motion, we have

$$v(F_1,\ |M_i| - \dot{M}_i) = \chi(M_i) - \omega_i = 0.$$

Then on applying Lemmas 2.1 and 2.2 and Theorem II5.1(iv), one can move, unite and finally cancel the fixed points of F_1 on $|M_i| - \dot{M}_i$ so that the modified F_1 is a fixed point free on $|M_i|$ and possesses the property S on $|M_i|$.

3) M_i is a 2–dimensionally connected complex, and g has at least one fixed point, say $a_1 = g(a_1)$, in $\dot{M}_i$. On applying Lemms 2.1 and 2.2, one can modify F_1 in $|M_i| - \dot{M}_i$ to move the fixed points in $|M_i| - \dot{M}_i$ to $(|M_i| - \dot{M}_i) \cap U(a_1, \varepsilon)$. Then on applying Lemma 2.1 to F_1 in $U(a_1, \varepsilon) \cap |M_i|$, one can unite all these fixed points onto a_1, so that the modified F_1 is free from fixed point in $|M_i| - \dot{M}_i$, $F_1(a_j) = g(a_j)$, $\forall a_j \in \dot{M}_i$, and F_1 possesses the property S on $|M_i|$.

Denote the totality of the modified F_1 for all the branches M_i by the mapping G: $|K| \to |K|$. This self-mapping G is what is desired. □

4. $\#\Phi(\langle id\rangle)$ of non-2-dimensionally-connected complex

As usual, let K be a connected finite simplicial complex. To generalize the concept of the fixed point classes of a self-mapping of $|K|$, we introduce in the present section the fixed point classes of a mapping of an open subset U of $|K|$ into $|K|$. We shall state without proof two basic properties of the new concept and use them to prove the conclusion $\#\Phi(\langle id\rangle) \geqslant \#\varphi(K)$ of Theorem 4.4, which together with Theorem 3.7 completes the proof of Theorem 3.5.

4.1 Definition. Let K be a connected finite simplicial complex, U an open subset of $|K|$, and $f: U \to |K|$ a mapping with compact fixed point set $\Phi(f)$. Two fixed points a and b of f are said to belong to the same fixed point subclass of f in U, when there exists *in* U a path p from a to b such that

$$f \circ p \simeq p \text{ on } |K|,$$

where $f \circ p$ is not necessarily in U.

When f is a self-mapping of $|K|$, and U a proper subset of $|K|$, we have fixed point class of f as well as fixed point subclass of $f \mid U$ in U. We shall also call the latter simply the *fixed point subclass of* f *in* U.

Let us note that when f is a self-mapping of $|K|$ and U a proper subset of $|K|$, two fixed points of the same fixed point subclass of f in U must belong to the same fixed point class of f, while two fixed points in U of the same fixed point class of f may belong to different fixed point subclasses of f in U. This justifies the use of the term fixed point subclass in Definition 4.1.

Since in both definitions of fixed point class and fixed point subclass the homotopies $\simeq$ considered are always on $|K|$, for brevity in the present section we shall mean by $\simeq$ or $\not\simeq$ always to $\simeq$ or $\not\simeq$ on $|K|$.

Replacing the self-mapping of a polyhedron in Definition II5.2 by $f: U \to |K|$ as described in the last definition, and U in II5.2 by V, we have the definition of the *index of a fixed point subclass of* f *in* U. An *essential fixed point subclass* is defined again as a fixed point subclass with its index not a zero. Just as Theorem II5.5 and Theorems II6.1 and II6.2, we have respectively the following two theorems.

4.2 Theorem. *Let U be an open subset of $|K|$ and $f\colon U\to|K|$ a mapping. Then the number of essential fixed point subclasses of f in U is finite.* □

4.3 Theorem. *Let U be an open subset of $|K|$, and $f_t\colon U\to|K|$, $t\in I$, a homotopy between the mapping f_0 and f_1, with $\bigcup_{t\in I}\Phi(f_t)$ compact in U. Then there exists a one-one correspondence between the essential fixed point subclasses of f_0 in U and those of f_1 in U. Moreover the corresponding subclasses have the same indices.* □

The following theorem together with Theorem 3.7 will yield Theorem 3.5.

4.4 Theorem. *Let K be a non-2-dimensionally-connected complex with $k(>1)$ branches and $f\simeq id$ a self-mapping of $|K|$. Then there exists a good star motion $\tilde{g}$ of the welding set $\dot{M}(K)$ of K such that*

$$\#\Phi(\tilde{g})\leqslant\#\Phi(f),$$

and hence

$$\#\varphi(K)\leqslant\#\Phi(\langle id\rangle).$$

Proof. If f has at least one fixed point in $|M_i|-\dot{M}_i, i=1,2,\cdots,k$, take an arbitrary point in $\dot{M}_i$ and regard it as a specified point. The union of these specified points and the fixed points of f in the welding set $\dot{M}(K)$ will be made the fixed point set $\Phi(\tilde{g})$ of the good star motion $\tilde{g}$ to be constructed. Denote the complement of this union in $\dot{M}(K)$ by $C=\{c_j\colon j=1,2,\cdots,h\}$. The construction of $\tilde{g}$ is simply to assign to every c_j an appropriate branch of K. It consists of the following three steps.

Step I. Take a sufficiently small number $\delta>0$ such that on $|K|$ all the neighborhoods $\overline{U}(c_j,\delta)$ of c_j, $j=1,2,\cdots,h$, are disjoint, every $\overline{U}(c_j,\delta)\subseteq\text{st } c_j$, and every $f(c_j)\notin\overline{U}(c_j,\delta)$.

The hypothesis $f\in\langle id\rangle$ in our theorem $\Rightarrow$ the existence of a homotopy f_t:

$$f_t\colon id\simeq f\colon |K|\to|K|.$$

For every c_j, denote by ${}^jp$ the path from c_j to $f(c_j)$ defined as follows

$${}^jp(t)=f_t(c_j),\ t\in I.$$

4.5 Lemma. *For every $c_j\in C$, There exists a path ${}^jq$ on $|K|$ and a point $b_j\in\partial U(c_j,\delta)$ such that*

(i) ${}^jp \doteqdot {}^jq$,

(ii) ${}^jq_0^{\frac{1}{2}} = [c_j, b_j]$,

(iii) *when* ${}^jq(s) \in U(c_j, \delta)$ *for some value* $s > \frac{1}{2}$, *then*

$$ {}^jq_0^s \not\doteqdot [c_j, {}^jq(s)]. $$

The expressions on the left in (ii) *and* (iii) *stand for subpaths of* ${}^jq$ (*Definition* A2.3), *while those on the right represent line segments.*

Remark. On replacing the unnecessary roundabout part at the beginning of ${}^jp$ by the line segment $[c_j, b_j]$ $(= {}^jq^{\frac{1}{2}})$, there results ${}^jq$ from ${}^jq$ (see Figure 11 and the proof of Lemma4.5). The point b_j will play an important role in the definition of g and $\tilde{g}$ at the beginning of Step III.

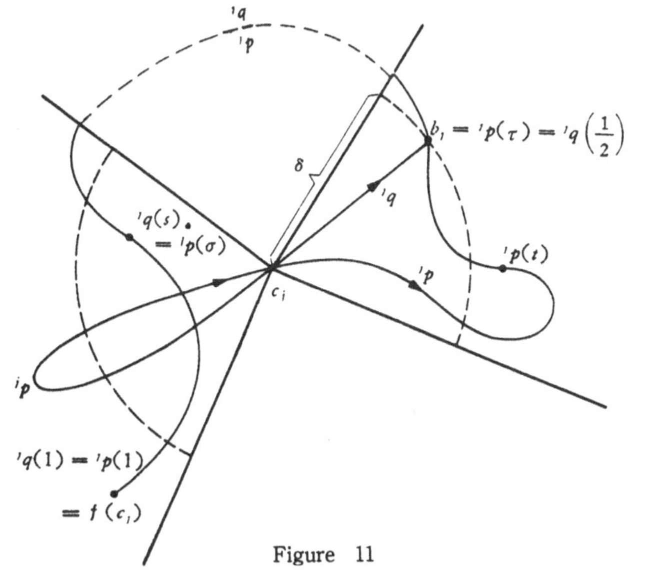

Figure 11

Proof. In order to simplify the notation, we shall omit the superscripts and the subscripts j. Let

$$ \tau = \sup\{t : t \in I,\ p(t) \in \overline{U}(c, \delta) \text{ and } p_0^t \doteqdot [c, p(t)] \text{ on } |K|\}, $$

$$ b = p(\tau). $$

It is not difficult to see that $\tau \in (0,1)$ and $b \in \partial U(c, \delta) \subseteq \text{st } c$.

Since p is continuous and st$c$ is open, there exists $\varepsilon > 0$ such that $p(s) \in \text{st} c$ for $|s - \tau| < \varepsilon$. From the definition of τ, there exists also $t \in (\tau - \varepsilon, \tau]$ such that $p_0^t \doteqdot [c, p(t)]$. As p_t^τ is a path in the simply connected region st c, we have $p_t^\tau \doteqdot [p(t), c, p(\tau)]$. Hence

$$ p_0^\tau \doteqdot p_0^t p_t^\tau \doteqdot [c, p(t)]\,[p(t), c, p(\tau)] \doteqdot [c, p(\tau)] = [c, b]. $$

Take $q=[c,b]p_\tau^1$. (ii) holds obviously.
(i) follows from the following

$$p \simeq p_0^\tau p_\tau^1 \simeq [c,b]p_\tau^1 = q.$$

To justify (iii), suppose $q(s) \in U(c,\delta)$ for a certain $s>\frac{1}{2}$. Set

$$\sigma=\tau+(1-\tau)(2s-1).$$

Then

$$p(\sigma)=q(s) \in U(c,\delta),$$

and

$$q_0^s \simeq [c,b]\, q_{1/2}^s \simeq [c,b]p_\tau^\sigma \simeq p_0^\tau p_\tau^\sigma \simeq p_0^\sigma.$$

Since $\sigma>\tau$, from the definition of τ

$$p_0^\sigma \not\simeq [c,p(\sigma)],$$

and hence

$$q_0^s \not\simeq [c,\ q(s)]. \qquad \square$$

Step II. Consider the following self-mapping $\bar f: |K|\to|K|$:

$$\bar f(x)=\begin{cases} x, & \text{when } x\in|K|-\bigcup_{j=1}^{h} U(c_j,\delta);\\ (\frac{2}{\delta}d(x,c_j)-1)x+(2-\frac{2}{\delta}d(x,c_j))c_j, & \\ \qquad \text{when } \frac{\delta}{2}\leqslant d(x,c_j)\leqslant\delta,\ j=1,2,\cdots,h; & \\ {}^jq(1-\frac{2}{\delta}d(x,c_j)), & \\ \qquad \text{when } 0\leqslant d(x,c_j)\leqslant\frac{\delta}{2},\ j=1,2,\cdots,h. & \end{cases} \tag{1}$$

We may say geometrically (see Definition 1.2) that $\bar f$ is obtained from *id* by attaching the path ${}^jp$ as a tail at c_j of $U(c_j,\delta)$, $j=1,2,\cdots,h$.

4.6 Lemma. $f\simeq\bar f: |K|\to|K|$ rel C.

Proof. Construct the two following homotopies F_t and G_t: $|K|\to|K|$.

$$F_t(x)=\begin{cases} f_t(x), & \text{when } x\in|K|-\bigcup_{j=1}^{h}U(c_j,\delta);\\ f_t((1-t)\bar{f}(x)+tx), & \\ \qquad \text{when } \frac{\delta}{2}\leqslant d(x,c_j)\leqslant\delta,\ j=1,2,\cdots,h; & \\ f_s((1-t)c_j+tx),\ \text{where } s=(1-\frac{2}{\delta}d(x,c_j))(1-t)+t, & \\ \qquad \text{when } 0\leqslant d(x,c_j)\leqslant\frac{\delta}{2},\ j=1,2,\cdots,h. & \end{cases}$$

$$G_t(x)=\begin{cases} \bar{f}_t(x), \text{when } x\in|K|-\bigcup_{j=1}^{h}U(c_j,\frac{\delta}{2}); \\ {}^j_t p(1-\frac{2}{\delta}d(x,c_j)), \\ \qquad \text{when } 0\leqslant d(x,c_j)\leqslant\frac{\delta}{2},\ j=1,2,\cdots,h; \end{cases}$$

where ${}^j_t p$ is the homotopy ${}^j p \simeq {}^j q$ between the paths in Lemma 4.5 (i).

It is not difficult to see that F_t and G_t are homotopies, $F_1=f$, $F_0=G_0$, $G_1=\bar{f}$, and the image of every c_j is preserved under both homotopies. Hence Lemma 4.6 holds.】

From (1), the fixed points of $\bar{f}: |K|\to|K|$ are of three different types.

Type 1. All points in $|K|-\bigcup_{j=1}^{h}U(c_j,\delta)$ are fixed points of $\bar{f}$.

Type 2. For each j, $\bar{f}$ has just one fixed point $d_j\in(c_j,b_j)$ in $U(c_j,\frac{\delta}{2})-\bar{U}(c_j,\frac{\delta}{4})$.

Type 3. For the fixed points of $\bar{f}$ in each $U(c_j,\frac{\delta}{4})$, we shall study first, in particular, the fixed points of $\bar{f}$ in a branch M_i of K with $\dot{M}_i\subseteq C$.

4.7 Lemma. *Let M_i be a branch of K with $\dot{M}_i\subseteq C$, and denote the set of fixed points of the first type of $\bar{f}$ in $|M_i|$ by*

$$m_i=|M_i|-\bigcup_j U(c_j,\delta).$$

Then (cf. Figure 12),

(i) *m_i and the set of fixed points of the second type of $\bar{f}$ in $|M_i|-\dot{M}_i$ belong to the same fixed point subclass of $\bar{f}$ in $|M_i|-\dot{M}_i$;*

(ii) *every fixed point of the third type of $\bar{f}$ in $|M_i|-\dot{M}_i$ does not belong to the same fixed point subclass of $\bar{f}$ in $|M_i|-\dot{M}_i$ which contains m_i;*

(iii) $v(\bar{f}, |M_i| - \bigcup_j \overline{U}(c_j, \frac{\delta}{4})) = 0$.

Proof. As $\dot{M}_i \subseteq C$, $\bar{f}$ has no fixed point in $\dot{M}_i$. Hence to $\bar{f}: |M_i| - \dot{M}_i \to |K|$ Theorem II5.1 can be applied and we can speak of the fixed point subclasses of $\bar{f}$ in $|M_i| - \dot{M}_i$.

(i) Consider the fixed point $d_j \in |M_i|$ of the second type of $\bar{f}$. Now $b_j \in m_i$, the segment $[b_j, d_j]$ is in $|M_i| - \dot{M}_i$, and from (1) we have

$$\bar{f} \circ [d_j, b_j] \simeq [d_j, c_j, b_j] \simeq [d_j, b_j].$$

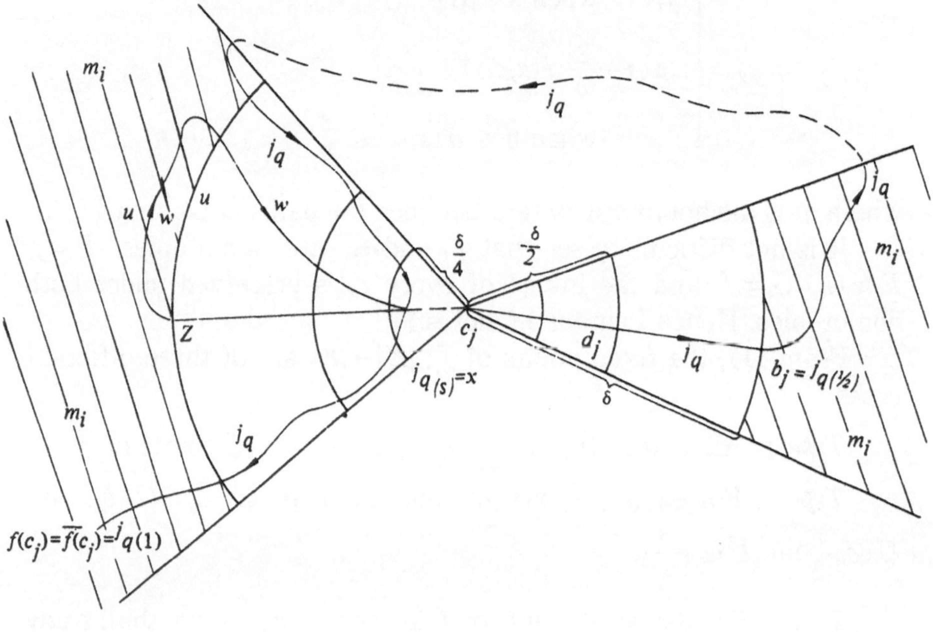

Figure 12

By Definition 4.1, the conclusion desired is obtained.

(ii) Let $x \in U(c_j, \frac{\delta}{4})$ be a fixed point of $\bar{f}$. From (1) there exists a value $s > 1/2$ such that $x = {}^jq(s)$. If z is the point of intersection of $\partial U(c_j, \delta)$ with the ray from c_j to x (Figure 12), then $z \in m_i$. (1) and Lemma 4.5 (iii) lead to

$$\bar{f} \circ [z, x] \simeq [z, c_j]\, {}^jq_0^s \not\simeq [z, c_j]\, [c_j, {}^jq(s)] \simeq [z, x]. \tag{2}$$

In order to prove that z and x do not belong to the same fixed point subclass of $\bar{f}$ in $|M_i| - \dot{M}_i$, let w be any path from z to x in $|M_i| - \dot{M}_i$. Radially project the part of w outside of $U(c_j, \delta)$ into $\partial U(c_j, \delta)$. From c_j, we obtain on m_i a closed path u from z to z (see Figure 12) such that

$$w \doteqdot u[z,x] \text{ on } |M_i| - \dot{M}_i. \tag{3}$$

From (2) and (3), we obtain

$$\bar{f}\circ w \doteqdot \bar{f}\circ(u[z,x]) = (\bar{f}\circ u)(\bar{f}\circ[z,x])$$
$$= u(\bar{f}\circ[z,x]) \not\doteqdot u[z,x] \doteqdot w.$$

By Definition 4.1 the conclusion desired is obtained.

(iii) From the definition of C, the hypothesis $\dot{M}_i \subseteq C$ implies that f has no fixed point in $|M_i|$, and thus f has no essential fixed point subclass in $|M_i| - \dot{M}_i$. According to Lemma 4.6 and Theorem 4.3, $\bar{f}$ has no essential fixed point subclass in $|M_i| - \dot{M}_i$, i.e., the index of every fixed point subclass of $\bar{f}$ in $|M_i| - \dot{M}_i$ is zero. However, (i) and (ii) of our lemma show that all the fixed points of $\bar{f}$ in $|M_i| - \bigcup_j \overline{U}(c_j, \frac{\delta}{4})$ constitute just one fixed point subclass of $\bar{f}$ in $|M_i| - \dot{M}_i$. Hence

$$v(\bar{f}, |M_i| - \bigcup_j \overline{U}(c_j, \tfrac{\delta}{4})) = 0.]$$

Step III. Define the mapping $\bar{g}: |K| \to |K|$ by the following

$$\bar{g}(x) = \begin{cases} \bar{f}(x), & \text{when } x \in |K| - \bigcup_j U(c_j, \frac{\delta}{4}); \\ b_j, & \text{when } x \in \overline{U}(c_j, \frac{\delta}{4}),\ \forall c_j \in C. \end{cases} \tag{4}$$

Geometrically speaking, $\bar{g}$ is obtained from $\bar{f}$ by cutting short the tails of $\bar{f}$ at c_j. Obviously $\bar{g}$ possesses the property S on $|K|$.

Let $g = \bar{g}|\dot{M}(K): \dot{M}(K) \to |K|$. From (4), g is the star mapping (Definition 3.3) as follows:

$$\begin{cases} g(a) = a, \ \forall a \in \dot{M}(K) - C; \\ g(c_j) = b_j, \ \forall c_j \in C. \end{cases} \tag{5}$$

4.8 Lemma. *Let $\tilde{g}$ be the star motion induced by g in* (5). *Then*

$$\#\Phi(\tilde{g}) \leqslant \#\Phi(f),$$

and $\tilde{g}$ is a good star motion (Definition 3.4).

Proof. From (5) there follows at once $\Phi(\tilde{g}) = \dot{M}(K) - C$. The definition of C implies $\#\Phi(\tilde{g}) \leqslant \#\Phi(f)$. It remains to show that $\tilde{g}$ is a good star motion.

Suppose M_i is a branch of K, and $\bar{g}$ has no fixed point on the welding set $\dot{M}_i$. We will verify that under $\bar{g}$, exactly $\chi(M_i)$ of the points of $\dot{M}_i$ correspond to branches of K other than M_i, or in other words, to show that if under $\bar{g}$, ω_i of the points of $\dot{M}_i$ corresponds to branches of K other than M_i, then $\omega_i = \chi(M_i)$.

By virtue of Lemma 3.6 and the fact that $\bar{g}$ has no fixed point in $\bar{U}(c_j, \frac{\delta}{4})$ for $\forall c_j \in \dot{M}_i$, we note

$$\chi(M_i) - \omega_i = v(\bar{g}, |M_i| - \dot{M}_i) = v(\bar{g}, |M_i| - \bigcup_j \overline{U}(c_j, \frac{\delta}{4})). \qquad (6)$$

Since $\bar{g} = \bar{f}$ in $|K| - \bigcup_j \overline{U}(c_j, \frac{\delta}{4})$, we have

$$v(\bar{g}, |M_i| - \bigcup_j \overline{U}(c_j, \frac{\delta}{4})) = v(\bar{f}, |M_i| - \bigcup_j \overline{U}(c_j, \frac{\delta}{4})). \qquad (7)$$

Now $\dot{M}_i \subseteq C$, and hence from (6), (7) and Lemma 4.7 (iii) there results $\omega_i = \chi(M_i)$. □

In the three steps above the proof of Theorem 4.4 has been completed. □

5. A sufficient condition for $\# \Phi(\langle f \rangle) = N(f)$[1)]

In the preceding three sections we dealt with the least number of fixed points of the mapping class of identity mapping. Now we turn to the mapping class $\langle f \rangle$ of a general self-mapping f, and shall give a condition on K under which $\# \Phi(\langle f \rangle) = N(f)$ (Theorem 5.3). The procedure in our proof is to move and unite the fixed points of f step by step, so that ultimately all the fixed points of every essential fixed point class of f are replaced by a single fixed point, and those of every non-essential fixed point class are cancelled. Lemma 5.1 aims at moving the fixed points of a fixed point class, while Lemma 5.2 at uniting them.

In contrast to what we did in Lemma 2.1 or 2.2, we shall no longer assume the self-mapping in Lemma 5.1 or 5.2 to possess the property S anywhere in $|K|$. The self-mapping f will be modified first in a neighborhood of a segment (by means of Lemma 1.5 on line-fence homotopy and other devices) so that the lemmas in § 2 can be applied to the modified mapping. It is in this sense that the two lemmas in the present section are more effective than those in § 2.

---

1) See the footnote on p.55.

5.1 Lemma. *Let K be a connected finite simplicial complex, σ_1 and σ_2 its two maximal simplexes of dimensions greater than* 1, *point $a\in\sigma_1$, point $b\in\bar\sigma_1\cap\bar\sigma_2$ but not a vertex of K, and $f\colon|K|\to|K|$ a self-mapping of $|K|$ having the point a as its only fixed point in a neighborhood V of $[a,b]$. Then there exists for some number $\varepsilon>0$ and for any point a' of $U(b,\varepsilon)\cap\sigma_2$ a self-mapping $F\colon|K|\to|K|$ of $|K|$ with the following properties*:

(i) $F\simeq f$ rel $|K|-V$;

(ii) *F has only one fixed point a' in V*;

(iii) *if z is a fixed point of f in $|K|-V$, w any path from z to a on $|K|$ such that $f\circ w \fallingdotseq w$, then*

$$F\circ(w[a,b,a'])\fallingdotseq w[a,b,a'].$$

Here, $\fallingdotseq$ means of course $\fallingdotseq$ on $|K|$.

Remark 1. The present lemma is a generalization of Lemma 2.2. It aims at moving the fixed point a of f from σ_1 to σ_2 when f is not assumed to possess the property S on $[a,b]$. The proof will proceed in three steps: 1) to construct a homotopy on $[a,b]$; 2) to modify the mapping f in a neighborhood of $[a,b]$ by means of Lemma 1.5 on line-fence homotopy so that the modified mapping possesses the property S on $[a,b]$; and finally 3) to apply Lemma 2.2 to move the fixed point a to a' along $[a,b]$.

Proof. We proceed in the three following steps.

1) Construction of a homotopy on $[a,b]$. Take $\delta>0$ such that

$$U([a,b],\delta)\subseteq V\cap \operatorname{stcar} b, \tag{1}$$

$$f(\overline{U}(b,\delta))\cap\overline{U}(b,\delta)=\emptyset, \tag{2}$$

$$x,\ f(x)\in\sigma_1,\ \forall x\in\overline{U}(a,2\delta). \tag{3}$$

Let (see Figure 13)

$$\begin{aligned} &c=\partial U\left(a,\frac{\delta}{2}\right)\cap[a,b],\ c'=\partial U(a,2\delta)\cap[a,b],\\ &d=\partial U(b,\delta)\cap[a,b], \end{aligned} \tag{4}$$

Let $l\colon[a,c]\to\partial\sigma_1$ be a preassigned broken line from a vertex of the closed carrier Car b to the intersection of the ray $\overrightarrow{cf(c)}$ with $\partial\sigma_1$. Then define a mapping $g_1\colon[a,b]\to|K|$ by the following:

$$g_1(x)=\begin{cases} f(x), & \text{when } f(x)\notin\sigma_1;\\ \text{the intersection } \overrightarrow{xf(x)}\cap\partial\sigma_1, & \text{when } f(x)\in\bar\sigma_1, x\in[c,d];\\ \text{the intersection } \overrightarrow{df(x)}\cap\partial\sigma_1, & \text{when } f(x)\in\bar\sigma_1,\ x\in[d,b];\\ l(x), & \text{when } x\in[a,c]. \end{cases}$$

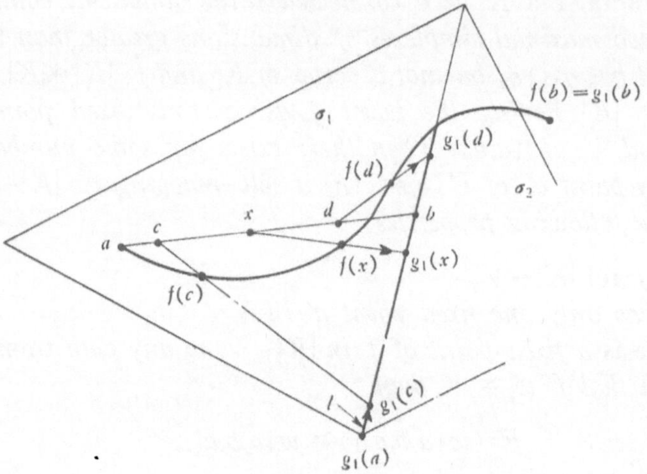

Figure 13 f and g_1

The g_1–image may contain also the point b. With a slight modification of g_1, we obtain a mapping $g_2: [a,b] \to |K|$ such that the g_2–image does not contain the point b. Let $G = |K - \sigma_1| \cap U(b,\delta)$. Then $g_1^{-1}(G)$ consists of a countable number of open intervals on $[a, b]$, and there are among them only a finite number containing points of $g_1^{-1}(b)$, say (a_j, b_j), $j=1,2,\cdots,n$. Denote by ∂G the boundary of G in $|K-\sigma_1|$. Then ∂G contains obviously $g_1(a_j)$ and $g_1(b_j)$, $j=1,2,\cdots, n$. Since b is not a vertex of K, ∂G is pathwise connected. Let p_j be

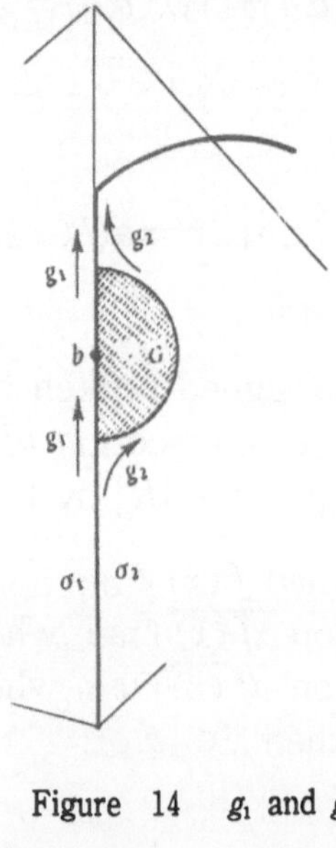

Figure 14 g_1 and g_2

a path on ∂G from $g_1(a_j)$ to $g_1(b_j)$. Since $\bar{G}$ is contractible, $g_1\circ[a_j,b_j]\simeq p_j$ on $\bar{G}$. Replace $g_1 \mid [a_j,b_j]$ with $g_2\circ[a_j,b_j]=p_j$, $j=1,2,\cdots,n$, to obtain the mapping $g_2:[a,b]\to|K|$ we have $g_1\simeq g_2$, and $b\notin g_2([a,b])$ (see Figure 14).

Set $H_1(x,t)=(1-t)f(x)+tg_1(x)$, $\forall x\in[a,b]$, $t\in I$, which is a homotopy joining $f|[a,b]$ and g_1. Let $H_2(x,t)$ be the homotopy joining g_1 and g_2 as described above, and $H_3(x,t)=g_2((1-t)x+ta)$, $\forall x\in[a,b]$, $t\in I$, the homotopy joining g_2 and the constant mapping $[a,b]\to g_2(a)$. From the definitions, one sees that

$$H_i(x,t)\neq x,\ \forall x\in[c,b],\ t\in I, \tag{5}$$

$$\text{Car } H_i(x,t)\cap\text{Car } x\neq\emptyset,\ \forall x\in[a,c'],\ t\in I. \tag{6}$$

Let the product of the three homotopies H_1, H_2 and H_3 (Definition A3.3) be denoted by $H:[a,b]\times I\to|K|$. From (5) and (6) there follow

$$H(x,t)\neq x,\ \forall x\in[c,b],\ t\in I;$$

$$\text{Car } H(x,t)\cap\text{Car } x\neq\emptyset,\ \forall x\in[a,c'],\ t\in I.$$

2) On applying Lemma 1.5, we obtain from the mapping f and the homotopy H on $[a,b]$ a mapping f': $|K|\to|K|$ such that $f'\simeq f$ rel $|K|-W([a,b],\eta)$, f' has no fixed point in $W([a,b],\eta,[c,b])$, f' possesses the property S on both $W([a,b],\eta,[a,c'])$ and $[a,b]$. From (4), one sees that, for sufficiently small η,

$$W([a,b],\eta,[a,c])\subseteq U(a,\delta), \tag{7}$$

$$W([a,b],\eta,[a,c'])\supseteq W([a,b],\eta)\cap U(a,\delta). \tag{8}$$

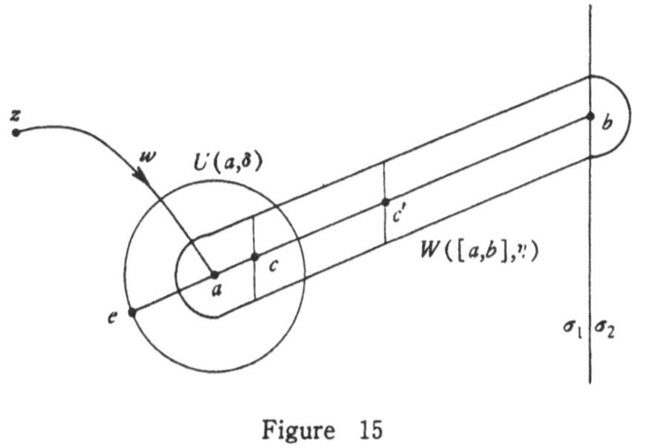

Figure 15

From (1), f has only one fixed point in $U(a,\delta)$, i.e. the point a. Now from (7), all the fixed points of f' in $W([a,b],\eta)$ lie in $U(a,\delta)$, and from (3) and (8), f' possesses the property S on $U(a,\delta)$.

Thus by applying Lemma 2.1, one can unite all the fixed points of f' in $U(a,\delta)$ onto the point a.

3) From the discussion above, we may consider f' to have only the single fixed point a in $U(a,\delta)$ and to possess the property S on $[a,b]$. From Lemma 2.2, for some $\varepsilon>0$ and for any given point $a'\in U(b,\varepsilon)\cap\sigma_2$, there exists a mapping $F\colon |K|\to|K|$ with the properties (i) and (ii) in our present Lemma.

We still need to prove that F possesses also the property (iii). Let z and w be given as in the hypothesis of (iii), and e be the intersection of the ray $\overrightarrow{ba}$ with $\partial W([a,b],\delta)$. Since $F\simeq f$ rel $\{z,e\}$, there follows $f\circ(w[a,e])\doteqdot F\circ(w[a,e])$. Since f possesses the property S on $[a,e]$, F possesses the property S on $[e,a,b,a']$, and

$$f(a)=a,\ F(a')=a',$$

there follows

$$(f\circ[a,e])(F\circ[e,a,b,a'])\doteqdot[a,e][e,a,b,a']\doteqdot[a,b,a'].$$

Hence

$$\begin{aligned}F\circ(w[a,b,a'])&\doteqdot(F\circ(w[a,e]))(F\circ[e,a,b,a'])\\&\doteqdot(f\circ(w[a,e]))(F\circ[e,a,b,a'])\\&\doteqdot(f\circ w)(f\circ[a,e])(F\circ[e,a,b,a'])\doteqdot w[a,b,a'].\end{aligned}$$

□

5.2 Lemma. *Let K be a connected finite simplicial complex, σ one of its maximal simplex of dimension greater than 2, and points a, $b\in\sigma$. If $f\colon|K|\to|K|$, a self-mapping of $|K|$, has just two fixed points a and b in a neighborhood V of $[a,b]$ and $f\circ[a,b]\doteqdot[a,b]$, then there exists a self-mapping $F\colon|K|\to|K|$ such that*

(i) $F\simeq f$ rel $\{|K|-V\}$;

(ii) *F has only one fixed point a in V;*

(iii) *if z is a fixed point of f in $|K|-V$, and w any path from z to a such that $f\circ w\doteqdot w$, then*

$$F\circ w\simeq w.$$

Remark 2. The present lemma is an extension of Lemma 2.1. It aims at uniting under certain condition two fixed points a and b of the same fixed point class. The proof proceeds also in three steps. The first two steps are the same as those in Remark 1, while the third is to apply Lemma 2.1 to move the fixed point b to a along $[a,b]$.

Proof. The hypothesis $f\circ[a,b]\doteqdot[a,b]$ means the existence of a homotopy $H_0\colon[a,b]\times I\to|K|$ such that

$$H_0(x,0)=f(x),\ H_0(x,1)=x,\ \forall x\in[a,b];$$

$$H_0(a,t)=a,\ H_0(b,t)=b,\ \forall t \in I.$$

1) Construction of a homotopy on $[a,b]$. Take positive number $\delta<\frac{1}{2}d(a,b)$ such that (cf. Figure 16)

$$U([a,b],\delta) \subseteq V \cap \sigma, \tag{9}$$

$$f(x) \in \sigma,\ \forall x \in \overline{U}(a \cup b,\ 2\delta), \tag{10}$$

$$H_0(x,t) \in \sigma,\ \forall x \in [a,c'] \cup [d',b],\ t \in I, \tag{11}$$

where

$$c'=\partial U(a,2\delta) \cap [a,b],\ d'=\partial U(b,2\delta) \cap [a,b],$$
$$c=\partial U(a,\frac{\delta}{2}) \cap [a,b],\quad d=\partial U(b,\frac{\delta}{2}) \cap [a,b]. \tag{12}$$

Then define a mapping $g:[a,b] \to |K|$ by the following:

$$g(x)=\begin{cases} f(x), & \text{when } f(x) \notin \sigma; \\ \text{the intersection } \overrightarrow{xf(x)} \cap \partial\sigma, & \text{when } f(x) \in \bar{\sigma},\ x \in [c,d]; \\ \text{the intersection } \overrightarrow{cf(c)} \cap \partial\sigma, & \text{when } x \in [a,c]; \\ \text{the intersection } \overrightarrow{df(d)} \cap \partial\sigma, & \text{when } x \in [d,b]. \end{cases}$$

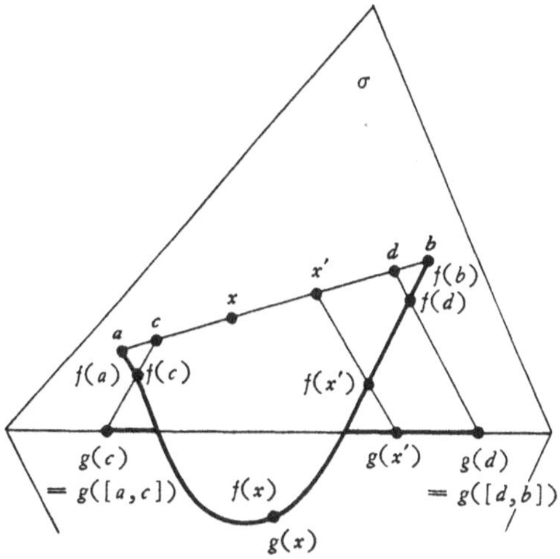

Figure 16 f and g

Let $H_1(x,t)=(1-t)f(x)+tg(x),\ \forall x \in [a,b],\ t \in I$. H_1 is a homotopy joining $f \mid [a,b]$ and g. From the definition of g and the fact that f is fixed point free in $[c,d]$, one sees

$$H_1(x,t) \neq x,\ \forall x \in [c,d],\ t \in I, \tag{13}$$

$$H_1(x,t) \in \bar{\sigma},\ \forall x \in [a,c'] \cup [d',b],\ t \in I. \tag{14}$$

Let $H':[a,b]\times I\to|K|$ be the homotopy joining g and id given by

$$H'(x,t)=\begin{cases}H(x,1-2t), & \text{when } 0\leqslant t\leqslant \frac{1}{2};\\ H_0(x,2t-1), & \text{when } \frac{1}{2}\leqslant t\leqslant 1.\end{cases}$$

From (11) and (14), there follows

$$H'(x,t)\in\bar{\sigma},\ \forall x\in[a,c']\cup[d',b],\ t\in I. \tag{15}$$

Let $H_3:[a,b]\times I\to|K|$ be a simplicial approximation of H' (from a certain triangulation of $[a,b]\times I$ to K). Set

$$H_2(x,t)=(1-t)g(x)+t\,H_3(x,0),\ \forall x\in[a,b],\ t\in I.$$

Then $H_2(x,t)$ is a homotopy joining $g(x)$ and $H_3(x,0)$. Since a triangulation of $[a,b]\times I$ is a 2-dimensional complex and the dimension of σ is >2, there follows $H_3(x,t)\in|K|-\sigma,\ \ \forall x\in[a,b],\ t\in I$; thus, since $g([a,b])\subseteq|K|-\sigma$, there follows

$$H_i(x,t)\neq x,\ \forall x\in[a,b],\ t\in I,\ i=2.3. \tag{16}$$

From (15) and $g(x)\in\partial\sigma,\ \forall x\in[a,c']\cup[d',b]$, there follows

$$H_i(x,t)\in\bar{\sigma},\ \forall x\in[a,c']\cup[d'b],\ t\in I,\ i=2,3. \tag{17}$$

Denote the product of the three homotopies H_1, H_2 and H_3 by $H:[a,b]\times I\to|K|$. From (13), (14), (16) and (17), there follow

$$H(x,t)\neq x,\ \forall x\in[c,d],\ t\in I,$$

$$H(x,t)\in\bar{\sigma},\ \forall x\in[a,c']\cup[d',b],\ t\in I.$$

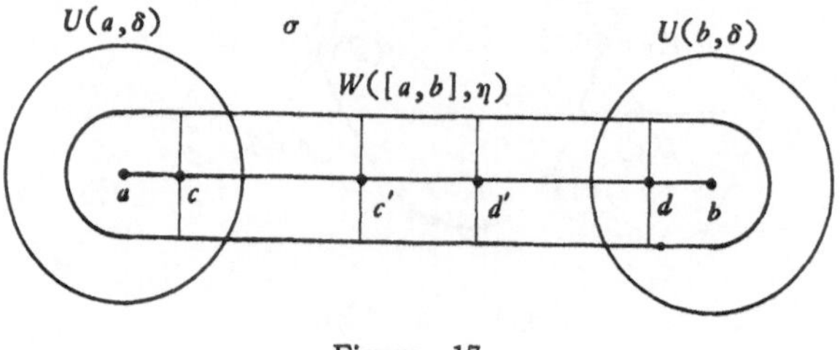

Figure 17

2) On applying Lemma 1.5, we obtain from the mapping f and the homotopy H a mapping f': $|K|\to|K|$ such that $f'\simeq f$ rel $|K|-W([a,b],\eta)$, f' has no fixed point in $W([a,b],\eta,[c,d])$, but possesses the property S on both $W([a,b],\eta,[a,c']\cup[d',b])$ and $[a,b]$. From (12), we take sufficiently small $\eta>0$ such that

$$W([a,b],\eta,[a,c]) \subseteq U(a,\delta), \tag{18}$$
$$W([a,b],\eta,[d,b]) \subseteq U(b,\delta),$$
$$W([a,b],\eta,[a,c']) \supseteq W([a,b],\eta) \cap U(a,\delta), \tag{19}$$
$$W([a,b],\eta,[d',b]) \supseteq W([a,b],\eta) \cap U(b,\delta).$$

From (9), the fixed points of f on $U([a,b],\delta)$ are just the points a and b. From (18), all the fixed points of f' on $W([a,b],\eta)$ lie in $U(a,\delta) \cup U(b,\delta)$. From (10) and (19), f' possesses the property S on $U(a,\delta) \cup U(b,\delta)$. Thus by applying Lemma 2.1, we can modify f' on $U(a,\delta)$ and $U(b,\delta)$ so that the terminal mapping has only one fixed point a on $U(a,\delta)$ and only one fixed point b on $U(b,\delta)$.

3) Since the terminal mapping just mentioned possesses the property S on $[a,b]$, we can apply Lemma 2.1 to moving the fixed point b toward a along $[a,b]$ and finally to uniting with a, and then we obtain a self-mapping F: $|K| \to |K|$ with the properties (i) and (ii) in our present lemma. The proof that F has the property (iii) is similar to the corresponding one in Lemma 5.1. □

5.3 Theorem (**Shi**). *Let K be a connected finite simplicial complex. If*

(i) *K is at least 3-dimensional, and*

(ii) *each of its vertex v is not a locally separating point, i.e., $\partial \operatorname{st} v$ is a connected sub-complex, then $\# \Phi(\langle f \rangle) = N(f)$ for every self-mapping f of $|K|$.*

Proof. 1) From Corollary II 6.3, we note $\# \Phi(\langle f \rangle) \geqslant N(f)$. Now it suffices to prove $\# \Phi(\langle f \rangle) \leqslant N(f)$.

For brevity, let us call a self-mapping *regular* when it has only a finite number of fixed points, each lying in the interior of a maximal simplex of K.

From Theorem C 2.5, for any self-mapping f of $|K|$, there is a regular self-mapping of $|K|$ in $\langle f \rangle$. Hence we need to prove only the equivalent assertion that if f is a regular self-mapping of our $|K|$ with $\# \Phi(f) > N(f)$, then there exists a regular self-mapping $g \in \langle f \rangle$, with $\# \Phi(g) \leqslant \# \Phi(f) - 1$.

2) For the regular self-mapping f in the above assertion, there are two possibilities. The first is the case that every pair of fixed points of f belongs to different fixed point classes. By virtue of the hypothesis that $\# \Phi(f) > N(f)$, at least one fixed point of f belongs to a non-essential fixed point class and its fixed point index is null. From Theorem II 5.1 (vi), one can remove this fixed point and obtain the self-mapping desired. Thus it remains only to consider the second

case that f has two fixed points a and b of the same fixed point class. We shall show below how to move and unite the fixed points a and b of f so as to obtain the desired g.

3) Now our f is a regular self-mapping of $|K|$ with $\#\Phi(f)>N(f)$, and its two fixed points a and b belong to the same fixed point class.

The condition (i) in our theorem implies that there exists a maximal simplex σ of dimension $\geqslant 3$ in K. The condition (ii) and the connectedness of K imply that K is 2-dimensionally connected. We may suppose $a \in \sigma$; for, if otherwise, by successive application of Lemma 5.1 one can move the fixed point a into σ, without having the fixed point class of b.

4) Because a and b belong to the same fixed point class, there exists a path w from a to b on $|K|$ such that $f\circ w \simeq w$ on $|K|$. On the basis of the condition (ii) in our theorem, in a construction similar to that of g_2 from g_1 in the proof of Lemma 5.1, one can obtain a path $w' \simeq w$ such that w' does not pass through any vertex of K (Figure 18). Consider the distances from the vertices of K to the image of w' and denote the shortest by $\eta>0$. Take a subdivision K' of K such that mesh $K'<\eta$, and finally obtain a broken-line w'' as simplicial approximation of w' satisfying $d(w''(t), w'(t))<\eta$, $\forall t \in I$. Denote by a_1 and b_1 the endpoints of w'', and set $l=[a,a_1]\,w''\,[b_1,b]$. The broken-line l so constructed is $\simeq w$, and does not pass through any vertex of K.

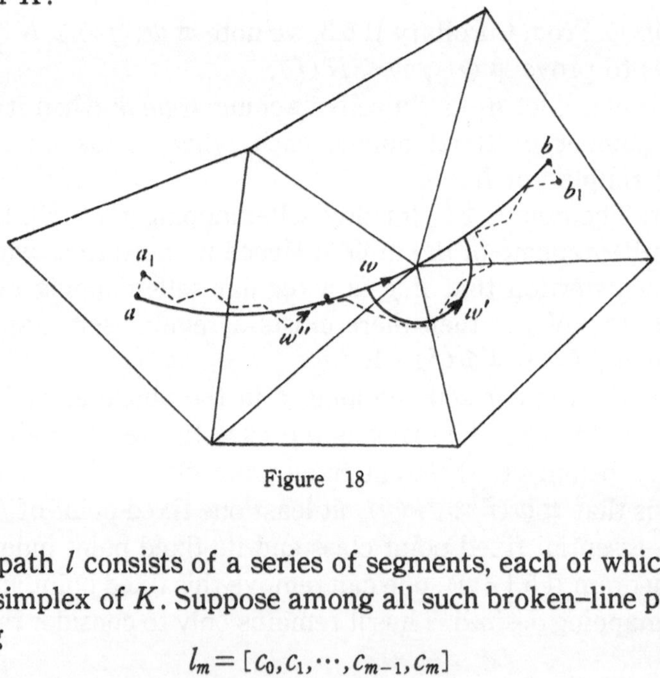

Figure 18

The path l consists of a series of segments, each of which lies in a closed simplex of K. Suppose among all such broken-line paths the following

$$l_m=[c_0,c_1,\cdots,c_{m-1},c_m]$$

is one of those with the least number of segments, where $c_0=a$ and $c_m=b$. Since f is regular, we take a sequence of maximal simplexes $\sigma_1=\sigma$, $\sigma_2,\cdots,\sigma_{m-1}$, $\sigma_m=\mathrm{car}\,b$ such that $[c_{j-1},c_j]\subseteq|\bar{\sigma}_j|$, $j=1,2,\cdots,m$. Moreover, every two succeeding simplexes in the sequence are different. In fact, if $\sigma_j=\sigma_{j+1}$, then $[c_{j-1},c_{j+1}]$ could be taken instead of $[c_{j-1},c_j,c_{j+1}]$; this is contrary to the hypothesis that l_m has the least number of segments.

$$l_m \simeq w \text{ implies } f\circ l_m \simeq l_m.$$

Since f is regular and the dimensions of the maximal simplexes are all $\geqslant 2$ by virtue of the condition (ii) in our theorem, we can modify slightly c_j in $\bar{\sigma}_j\cap\bar{\sigma}_{j+1}$, $j=1,2,\cdots,m-1$, so that f has no other fixed point on l_m besides the fixed endpoints a and b.

5) Set $f_m=f$ and $d_m=c_m$. On applying Lemma 5.1 to moving the fixed point of f_m at d_m along $[d_m,c_{m-1}]$ to a point d_{m-1} of σ_{m-1}, one obtains a mapping $f_{m-1}\simeq f_m$: $|K|\to|K|$. Moreover the point d_{m-1} can be so chosen that f_m is fixed point free on $[d_{m-1},c_{m-2}]$, and hence f_{m-1} has no fixed point on the broken-line $l_{m-1}=[c_0,c_1,\cdots,c_{m-2}, d_{m-1}]$ besides the endpoints. From Lemma 5.1 (iii) there follows (Figure 19)

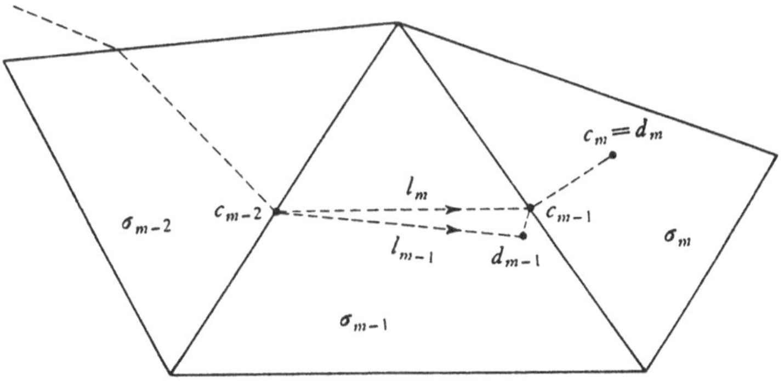

Figure 19

$$f_{m-1}\circ(l_m[d_m,\ c_{m-1},\ d_{m-1}])\simeq l_m[d_m,\ c_{m-1},\ d_{m-1}].$$

On the other hand,

$$\begin{aligned} l_m[d_m,\ c_{m-1},\ d_{m-1}] &\simeq [c_0,c_1,\cdots,c_{m-2},c_{m-1},d_m,c_{m-1},d_{m-1}] \\ &\simeq [c_0,c_1,\cdots,c_{m-2},c_{m-1}]=l_{m-1}, \end{aligned}$$

hence

$$f_{m-1}\circ l_{m-1}\simeq l_{m-1}.$$

Repeating the same method, we move the fixed point $d_{m-1}\in\sigma_{m-1}$ to certain point $d_{m-2}\in\sigma_{m-2}$ and obtain mapping f_{m-2}, and finally we obtain $f_1\simeq f_m=f$: $|K|\to|K|$. The net result is to have the fixed point

$d_m = b$ moved to the fixed point $d_1 \in \sigma_1 = \sigma$ of f_1; thus f_1 has no other fixed points in $l_1 = [c_0, d_1]$ than the two fixed endpoints and $f_1 \circ [c_0, d_1] \simeq [c_0, d_1]$.

Now by applying Lemma 5.2, we unite the fixed points d_1 with $c_0 = a$ of f_1, and obtain a mapping $g \in \langle f \rangle$, which has one less fixed point than f does. □

Theorem 5.3 together with Theorem III 5.1 give rise to the following

5.4 Theorem (A Converse of Lefschetz Fixed Point Theorem). *Let K be a connected finite simplicial complex and f a self-mapping of* $|K|$. *If K is at least 3-dimensional and has no locally separating vertex, and if* $J(f, x_0) = \pi_1(|K|, f(x_0))$, *and* $L(f) = 0$, *then there exists in* $\langle f \rangle$ a *fixed point free self-mapping g of* $|K|$. □

Another consequence of Theorem 5.3 is

5.5 Theorem. *The Nielsen number* $N(f)$ *of a self-mapping f equals the least number of fixed points among all self-mappings having the same homotopy type as f.*

Proof. Note that for every finite connected polyhedron K there always exists an at least 3-dimensional polyhedron M with $|M| \simeq |K|$, which has no locally separating vertex. (For example, $|M|$ is $|K| \times \underline{\sigma}$, where $\underline{\sigma}$ is a 3-dimensional simplex.) Then the conclusion follows Theorem 5.3. II (cf. Jiang Boju's book *Lectures on Nielsen Fixed Point Theory*, p.23, Theorem 6.4).

Chapter V

THE NUMBER $N(f;H)$ AND THE ROOT CLASSES

Based on [25] there has been further abvance in several directions. we shall present briefly two of these directions in the present chapter.

A. *The number* $N(f;H)$ [30]

1. Definitions and theorems under basic assumption

Let $(\tilde{X},p)$ be a k–leaved regular covering space (see definition preceding Theorem B3.11) of a connected finite polyhedron X, k being a natural number. Thus, for any $x_0 \in X$ and $\tilde{x}_0 \in p^{-1}(x_0)$, the group $H(X,x_0)=p_\pi(\pi_1(\tilde{X},\tilde{x}_0))$ is a normal subgroup of $\pi_1(X,x_0)$ and $\#(\pi_1(X,x_0)/H(X,x_0))=k$ (Corollary B1.6). If $\tilde{x}_1$, $\tilde{x}_2 \in p^{-1}(x_0)$, then from theorem B3.11 there exists a unique covering motion γ such that $\gamma(\tilde{x}_1)=\tilde{x}_2$. From Theorem B3.12, there follows that the group of covering motions $\mathscr{D}(\tilde{X},p)\approx \pi_1(X,x_0)/H(X,x_0)$ and consequently $\#\mathscr{D}(\tilde{X},p)=k$.

We consider only such self–mappings f of X that

$$f_\pi(H(X,x_0))\subseteq H(X,f(x_0)). \tag{1}$$

Let $f(x_0)=x_0'$. Then from Theorem B2.4, for any $\tilde{x}_0 \in p^{-1}(x_0)$ and any $\tilde{x}_0' \in p^{-1}(x_0')$, f has a unique lifting $\tilde{f}:(\tilde{X},\tilde{x}_0)\to(\tilde{X},\tilde{x}_0')$ such that

$$p\circ\tilde{f}=f\circ p; \tag{2}$$

and hence f has exactly k different liftings.

Throughout § 1, we shall make the *basic assumption* that (1) holds for the self–mapping f of X and the k–leaved regular covering space $(\tilde{X},p)$ considered. All propositions here will be based on this assumption, even if there is no explicit mention of it in their statements.

First, from discussion similar to that in II § 1, we have at once the following consequences on the fixed point sets $\Phi(f)$ and $\Phi(\tilde{f})$. If $\tilde{f}'$ is a lifting of the self-mapping f of X, then

$$p(\Phi(\tilde{f}))\subseteq\Phi(f);$$

if $\tilde{f}$ and $\tilde{f}'$ are two liftings of f and

$$p(\Phi(\tilde{f}))\cap p(\Phi(\tilde{f}'))\neq\emptyset,$$

then there exists a covering motion $\gamma\in\mathscr{D}(\tilde{X},p)$ such that

$$\tilde{f}'=\gamma\circ\tilde{f}\circ\gamma^{-1},$$

and hence

$$p(\Phi(\tilde{f}'))=p(\Phi(\tilde{f})).$$

For a given $\tilde{f}$, the set

$$\{\tilde{f}'=\gamma\circ\tilde{f}\circ\gamma^{-1}\colon \gamma\in\mathscr{D}(\tilde{X},p)\}$$

is called a *lifting H-class* of f, and is denoted by $[\tilde{f}]$. It is obvious that $[\tilde{f}']=[\tilde{f}]$ when $\tilde{f}'=\gamma\circ\tilde{f}\circ\gamma^{-1}$ for a certain $\gamma\in\mathscr{D}(\tilde{X},p)$. Call $p(\Phi(\tilde{f}))$ a *fixed point H-class* of f. Since it is independent of the choice of $\tilde{f}$ from $[\tilde{f}]$, we may denote it by $p(\Phi([\tilde{f}]))$. Each fixed point of f belongs to a fixed point H-class of f, and the set $\Phi(f)$ is separated into disjoint fixed point H-classes of f. Thus we have the following in correspondence with Theorem II 1.4:

1.1 Lemma. *Let $\tilde{f}$, $\tilde{f}'$,… be the liftings of the self-mapping f of X. Then among the fixed point sets $\Phi(f)$, $\Phi(\tilde{f})$, $\Phi(\tilde{f}')$,… the following hold:*

$$\Phi(f)=\bigcup_{\tilde{f}} p(\Phi(\tilde{f}))\ \textit{for all liftings}\ \tilde{f}\ \textit{of}\ f;$$

$$[\tilde{f}]=[\tilde{f}']\Rightarrow\ p(\Phi(\tilde{f}))=p(\Phi(\tilde{f}'));$$

$$[\tilde{f}]\neq[\tilde{f}']\Rightarrow\ p(\Phi(\tilde{f})\cap p(\Phi(\tilde{f}'))=\emptyset.\qquad\square$$

Because $\tilde{X}$ is a finite-leaved covering space of a connected finite polyhedron X, $\tilde{X}$ is also a connected finite polyhedron, and a lifting $\tilde{f}:\tilde{X}\to\tilde{X}$ has its Lefschetz number $L(\tilde{f})$. For $\tilde{f}'=\gamma\circ\tilde{f}\circ\gamma^{-1}$, $\gamma\in\mathscr{D}(\tilde{X},p)$, we have $L(\tilde{f}')=L(\tilde{f})$, since γ is a homeomorphism of $\tilde{X}$. Thus we can define by

$$L([\tilde{f}])=L(\tilde{f})$$

the *Lefschetz number of the lifting H-class* $[\tilde{f}]$.

Next, we proceed as in II § 3 to consider a homotopy $F: f_0 \simeq f_1$ between the self-mappings $f_0, f_1: X \to X$. The basic assumption on f_0 implies that f_1 satisfies the basic assumption, too. Let $\tilde{f}_0: \tilde{X} \to \tilde{X}$ be a lifting of f_0. From Theorem B2.5, F has a unique lifting $\tilde{F}$ with $\tilde{f}_0$ as the starting point. Let $\tilde{f}_1$ be the terminal point of $\tilde{F}$. Obviously $L(\tilde{f}_0) = L(\tilde{f}_1)$. In terms of the discussion in II § 3, a given homotopy F gives rise to an one–one correspondence between the lifting H–classes of f_0 and the lifting H–classes of f_1, and the Lefschetz numbers of the two corresponding lifting H–classes $L([\tilde{f}_0])$ and $L([\tilde{f}_1])$ obviously equal. This may be stated as the following lemma (cf. Theorem II6.1).

1.2 Lemma. *Let $F: f_0 \simeq f_1: X \to X$ be a homotopy and $\tilde{f}_0: \tilde{X} \to \tilde{X}$ any lifting of f_0. If the lifting $\tilde{f}_1: \tilde{X} \to \tilde{X}$ corresponds to $\tilde{f}_0$ under F (Definition* II3.1), *then $L([\tilde{f}_1]) = L([\tilde{f}_0])$.* □

We are now in a position to define the number $N(f;H)$ that has appeared in the title of Part A.

1.3 Definition. The number of the lifting H–classes of the self–mapping f of X, which have non–null Lefschetz numbers, is denoted by $N(f;H)$. It may be called *the Nielsen number of the H-class of f.*

1.4 Theorem. *Any self–mapping of X homotopic to f has at least $N(f;H)$ fixed points.*

Proof. Because of Lemma 1.2 and Definition 1.3, we need consider only f and its liftings. $L(\tilde{f}) \neq 0 \Rightarrow$ the fixed point H-class $p(\Phi(\tilde{f}))$ is not empty. Different fixed point H-classes are disjoint subsets of $\Phi(f)$. □

1.5 Corollary (Hirsch Theorem[1]). *Let X be a connected finite polyhedron, $(\tilde{X}, p)$ a two-leaved covering space of X, and f a self–mapping of X. Moreover, let*

(i) *f have liftings $\tilde{f}_i: \tilde{X} \to \tilde{X}$, $i = 1,2$;*
(ii) *for $a, b \in \tilde{X}$, $a \neq b$, and $p(a) = p(b)$, $\tilde{f}_i(a) \neq \tilde{f}_i(b)$, $i = 1,2$;*
(iii) *$L(\tilde{f}_i) \neq 0$, $i = 1,2$.*

Then every self-mapping of X homotopic to f has at least two fixed points.

1) G. Hirsch, Détermination d'un nombre minimum de points fixes pour certaines représentations, *Bull. Sci. Math.*, **64** (1940), 45—55.

Proof. For a two–leaved covering space $\tilde{X}$ of X, the subgroup $H(X,x_0)=p_\pi(\pi_1(\tilde{X},\tilde{x}_0))$ of $\pi_1(X,x_0)$ is of index 2 in $\pi_1(X,x_0)$ and hence is a normal subgroup of $\pi_1(X,x_0)$, and $\# \dfrac{\pi_1(X,x_0)}{H(X,x_0)}=2$. Consequently, $\mathscr{D}(\tilde{X},p)$ consists of two elements {identity,γ}, γ interchanging the two points of every fibre. From (i), f satisfies our basic assumption (1). From (ii),

$$\gamma\circ\tilde{f}_i\circ\gamma^{-1}(a)=\gamma\circ\tilde{f}_i(b)=\tilde{f}_i(a),\ i=1,2;$$

that is, $[\tilde{f}_i]$ consists of the single element $\tilde{f}_i$, and hence

$$[\tilde{f}_1]\neq[\tilde{f}_2].$$

From (iii), $N(f;H)=2$. □

1.6 Theorem. *If both H and H' are normal subgroups of $\pi_1(X,x_0)$ with $H'\subseteq H$ and both $N(f;H)$ and $N(f;H')$ are defined, then $N(f;H)\leqslant N(f;H')$. Consequently*

$$N(f;H)\leqslant N(f).$$ □

We shall leave the proof of this theorem to the reader and make only the following remark in the extreme cases: 1) H'=the identity element and 2) $H=\pi_1(X,x_0)$. In case 1), by definition we have $N(f;H')=N(f)$. In case 2), $N(f:\pi_1(X,x_0))$ is 0 or 1 according as $L(f)=0$ or $\neq 0$ respectively.

2. Examples

The present section is devoted to the description of the two examples in [30]. The space X in either example is a closed orientable manifold, and the self–mapping f of X is a self homeomorphism. To construct the f, we need the following elementary self homeomorphism of an annulus, called the *Dehn twist*[1]. Its explanation in Lickorish's words is as follows. "Let the 2–manifold M contain an annulus A, one of the boundary components of which is a simple closed curve c. There is a homeomorphism of A to itself, fixed on the boundary of A, which sends radial arcs onto arcs which spiral once (or several times) around A (see figure). This can be extended to a

1) W.B.R. Lickorish, Homeomorphisms of non–orientable manifolds, *Proc. Camb. Phil. Soc.*, **59** (1963), 307—317. See also J. Stillwell, *Classical Topology and Combinatorial Group Theory*, 1980, Springer–Verlag, p. 198.

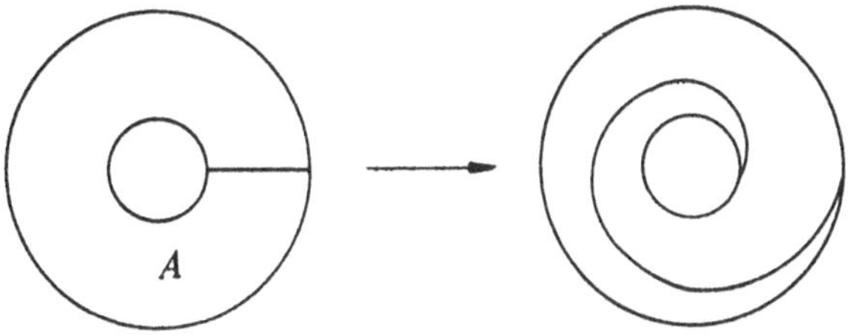

homeomorphism of M to itself, by the identity on $M-A$. Intuitively this homeomorphism can be thought of as the process of cutting M along c, twisting one of the new free ends, and then glueing together again." In the figure the outer circle c is twisted once in the counter clockwise sense .

In either of the two examples, the Lefschetz number $L(f)$ of the self–homeomorphism f of X is null, while by means of a two–leaved regular covering space $\tilde{X}$ we find $N(f;H)=2$. Hence from Corollary 1.5, every self–mapping homotopic to f has at least 2 fixed points.

Example 1. X is T_2 and the 2–leaved regular covering surface $\tilde{X}$ of X is T_3, where T_k is the closed orientable surface of genus k. Suppose that both surfaces T_2 and T_3 are smooth, symmetric and lie on a horizontal plane and that the curves $\alpha_i,\beta_i,\tilde{\alpha}_i,\tilde{\beta}_i$ and the three simple closed curves without label are all circles. See Figure 1.

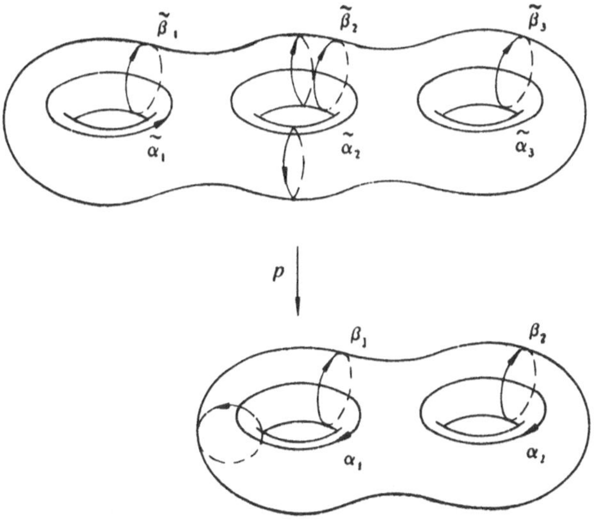

Figure 1

$\alpha_1,\beta_1,\alpha_2,\beta_2$ constitute a basis of the one–dimensional homology group $H_1(T_2)$ of T_2. The self–homeomorphism of T_2 to be constructed is $f = t_2 \circ t_1$, where t_1 and t_2 are the Dehn twists given as follows. The twist t_1 is to cut T_2 along β_1, to twist the left bank of the cut through 360° in the sense opposite to that of β_1, and then to glue together again. The twist t_2 is to cut T_2 along α_1, to twist the outer bank of the cut through 720° in the sense same as that of α_1, and then to glue together again. The effect of these two twists on the pair of generators α_1 and β_1 are given by the following two matrices respectively:

$$A_1 = \begin{pmatrix} 1 & -1 \\ 0 & 1 \end{pmatrix} \text{ and } A_2 = \begin{pmatrix} 1 & 0 \\ 2 & 1 \end{pmatrix},$$

and hence that of f is given by the product

$$A = A_1 A_2 = \begin{pmatrix} -1 & -1 \\ 2 & 1 \end{pmatrix}.$$

As f does nothing to the other pair of generators α_2 and β_2, its effect on them is given by the identity matrix E (2×2 matrix). The effect of f on the basis $\alpha_1, \beta_1, \alpha_2, \beta_2$ is given thus by

$$\left(\begin{array}{c|c} A & 0 \\ \hline 0 & E \end{array}\right)$$

the matrix of transformation of basis of $H_1(T_2)$ by $f_{1*}: H_1(T_2) \to H_1(T_2)$.

As f is a self-homeomorphism which preserves orientation, we have

$$f_{2*} = id\colon H_2(T_2) \to H_2(T_2).$$

Thus, as the alternative sum of the traces $\operatorname{tr}(f_{i*})$, $i = 0,1,2$, we have

$$L(f) = 1 - 2 + 1 = 0.$$

Next, let us see how to define the projection $p\colon T_3 \to T_2$ so that (T_3, p) is a 2-leaved covering of T_2. Call the vertical line through the center of the central hole of T_3 the axis of T_3. Cut T_3 into two halves along the two unlabeled simple closed curves. For each half, identify the two boundary curves by identifying every pair of points which are symmetric in the axis, and then take a homeomorphism from the resulting space onto T_2 such that it sends the curve of identification onto the unlabeled simple closed curve of T_2. This defines the required projection p. The group $\mathscr{D}(T_3, p)$ of covering motions consists of two elements id and γ, where γ denotes the rotation of T_3 through 180° around the axis.

Finally, from the self-mapping f of T_2 we shall construct first a lifting $\tilde{f}: T_3 \to T_3$ as the composition of two Dehn's twists: $\tilde{f}=t_2 \circ t_1$. The twist t_1 is to cut T_3 along $\tilde{\beta}_2$, to twist the left bank of the cut through 720° in the sense opposite to that of $\tilde{\beta}_2$, and then to glue together again. The twist t_2 is to cut T_3 along $\tilde{\alpha}_2$, to twist the outer bank of the cut through 360° in the sense same as that of $\tilde{\alpha}_2$, and then to glue together again. The effect of these two twists on the pair of generators $\tilde{\alpha}_2$ and $\tilde{\beta}_2$ is given by the following two matrices respectively:

$$B_1=\begin{pmatrix}1 & -2\\ 0 & 1\end{pmatrix} \text{ and } B_2=\begin{pmatrix}1 & 0\\ 1 & 1\end{pmatrix},$$

and hence that of $\tilde{f}$ is given by the product

$$B=B_1B_2=\begin{pmatrix}-1 & -2\\ 1 & 1\end{pmatrix}.$$

As $\tilde{f}$ does nothing to the pair $\tilde{\alpha}_1$ and $\tilde{\beta}_1$ and the pair $\tilde{\alpha}_3$ and $\tilde{\beta}_3$, the effect of $\tilde{f}$ on either pair is given by the identity matrix. The reader now can see easily

$$L(\tilde{f})=1-4+1=-2.$$

Let $\tilde{f}'=\gamma\circ\tilde{f}$. Then the effect of $\tilde{f}'$ on the basis of $H_1(T_3)$ is given by

$$\left(\begin{array}{c:c:c} E & 0 & 0\\ \hdashline 0 & B & 0\\ \hdashline 0 & 0 & E\end{array}\right)\cdot\left(\begin{array}{c:c:c} 0 & 0 & E\\ \hdashline 0 & E & 0\\ \hdashline E & 0 & 0\end{array}\right)=\left(\begin{array}{c:c:c} 0 & 0 & E\\ \hdashline 0 & B & 0\\ \hdashline E & 0 & 0\end{array}\right).$$

Thus $L(\tilde{f}')=1-0+1=2$.

It is then easy to see that we can apply Corollary 1.5 and obtain the desired result.

Example 2. T_2 in Example 1 is 2-dimensional. On modifying that example, we are going to give a self-homeomorphism h of an n-dimensional ($n \geqslant 3$) closed orientable manifold X such that $L(h)=0$ but every sely-mapping homotopic to h has at least 2 fixed points.

Let $g:S^{n-2}\to S^{n-2}$ be a self-homeomorphism of the $(n-2)$-dimensional sphere with $L(g)\neq 0$. Let $X=T_2\times S^{n-2}$ and $h=f\times g$ where f is the self-homeomorphism of T_2 given in Example 1. Then

$$h=f\times g: T_2\times S^{n-2}\to T_2\times S^{n-2}$$

is a self-homeomorphism of the n-dimensional manifold X, and from Theorem IV B6 in [2]

$$L(h)=L(f)\cdot L(g)=0.$$

Making use of the projection $p:T_3\to T_2$ and the liftings $\tilde{f}$ and $\tilde{f}'$ in Example 1, we find that $(T_3\times S^{n-2},\ p\times id)$ is a 2-leaved covering space of $T_2\times S^{n-2}$,

$$\tilde{h}=\tilde{f}\times g,\ \tilde{h}'=\tilde{f}'\times g\colon T_3\times S^{n-2}\to T_3\times S^{n-2}$$

are two liftings of h, and $L(\tilde{h})=L(\tilde{f})\cdot L(g)\neq 0$, $L(\tilde{h}')=L(\tilde{f}')$. $L(g)\neq 0$. Thus every self-mapping homotopic to f has at least two fixed points.

B. *The root classes*[17]

3. From fixed point classes of self-mapping to root classes of equation

Let the connected finite polyhedron X be a topological group[1)], 1 its identity element and x_* a given point. Then, for a self-mapping f of X, x is a root of the equation

$$f(x)=x_* \tag{1}$$

$\Leftrightarrow x$ is a root of the equation

$$x_*^{-1}\cdot f(x)=1 \tag{2}$$

$\Leftrightarrow x$ is a fixed point of the self-mapping

$$g(x)=x_*^{-1}\cdot f(x)\cdot x \tag{3}$$

of X. This shows that for a topological group X, the roots of the equation (1) and the fixed points of the self-mapping g of X are closedly related.

From this relationship, there naturally arises from the concept of fixed point class the concept of "*root class*". Let x_0 and x_1 be two fixed points of the self-mapping g of X. From Theorem II2.1, they are of the same fixed point class $\Leftrightarrow$ there exists in X a path c from x_0 to x_1 such that

$$c \mathrel{\dot{\simeq}} g\circ c; \tag{4}$$

i.e., there exists a homotopy G with fixed endpoints between c and $g\circ c$:

1) P.J. Hilton and S. Wylie, *Homology Theory*, Camb. Univ. Press, 1960, Definition 6.9.1.

$$G: I \times I \to X, \ (t,s) \mapsto G(t,s),$$

$G(t,0)=c(t)$, $G(t,1)=g\circ c(t)$, $G(0,s)=c(0)=x_0$, $G(1,s)=c(1)=x_1$. On the other hand, from (3) we have

$$f(x)=x_* \cdot g(x) \cdot x^{-1}. \tag{5}$$

Let

$$F: I \times I \to X, \ (t,s) \mapsto F(t,s)=x_* \cdot G(t,s) \cdot c(t)^{-1},$$

where $c(t)^{-1}$ denotes the inverse of the point $c(t)$ of X but not the inverse c^{-1} of the path c. Thus,

$$F(t,0)=x_*, \ F(t,1)=x_* \cdot g\circ c(t) \cdot c(t)^{-1}=f\circ c(t),$$

$$F(0,s)=x_*, \ F(1,s)=x_*.$$

This proves the part "$\Rightarrow$" of the following proposition.

Proposition. *Let x_0 and x_1 be two roots of the equation $f(x)=x_*$. Then, they are fixed points of the same fixed point class of the self-mapping $g(x)=x_*^{-1} \cdot g(x) \cdot x$ of X $\Leftrightarrow$ there exists in X a path c from x_0 to x_1 such that*

$$f\circ c \simeq e_*,$$

where e_* is the point path $I \to x_*$.

The proof of the part "$\Leftarrow$" is similar.

When the statement after "$\Leftrightarrow$" in our Proposition is taken as the definition of root class, then for a topological group X, a fixed point class of g in (3) is a root class of the equation (1), and conversely.

In the discussion above the space X is assumed to be a topological group. Hereafter we explain briefly some of the main results of [17] *free from this restriction.* We shall consider only arcwise connected topological spaces even when we do not mention this explicitly.

3.1 Definition. Let f be a mapping from an arcwise connected topological space Y to an arcwise connected topological space X, and x_* a given point in X. Denote *the set of all roots of the equation*

$$f(y)=x_*$$

by $\Gamma(f, x_*)$, i.e., $\Gamma(f, x_*)=f^{-1}(x_*)$. Two roots y_0 and y_1 are said to be of the *same class,* when there exists a path c in Y from y_0 to y_1 such that

$$f\circ c \simeq e_*, \text{ or } \langle f\circ c\rangle = e,$$

where $e_*:I\to x_*$ is the point path and e the identity element of $\pi_1(X, x_*)$.

Clearly, the relation of belonging to the same root class is an equivalence relation and hence $\Gamma(f, x_*)$ is separated into disjoint subsets. Each of these subsets is called a *root class*. (Please note that a root class is never an empty set.) Denote *the set of root classes* by $\Gamma'(f, x_*)$.

3.2 Lemma. *$\Gamma(f, x_*)=f^{-1}(x_*)$ is closed in Y, if X is a T_1 space.*

Proof. From the hypothesis, $\{x_*\}$ is a closed set in X, and so is $\Gamma(f, x_*)=f^{-1}(x_*)$ is a closed set in Y. □

3.3 Lemma. *If Y is locally arcwise connected and X is locally simply connected, then any root class $\mathcal{R}\in\Gamma'(f, x_*)$ is an open subset of $\Gamma(f, x_*)$.*

Proof. Let $y\in\mathcal{R}$. Take a neighborhood V of x_* in X such that $\langle c\rangle=\langle e_*\rangle$ for any loop c at x_* in V. As $f^{-1}(V)$ is a neighborhood of y, it contains an arcwise connected neighborhood U of y. We claim: if $y'\in U\cap\Gamma(f, x_*)$, then $y'\in\mathcal{R}$. In fact, in the arcwise connected U there exists a path d from y to y'; the f–image $f\circ d$ is in V and is a loop at x_*, and hence $\langle f\circ d\rangle=\langle e_*\rangle$. This shows $U\cap\Gamma(f, x_*)\subseteq\mathcal{R}$. Thus $\mathcal{R}$ is open in $\Gamma(f, x_*)$. □

3.4 Theorem. *If Y is compact, locally arcwise connected and X is locally simply connected T_1–space, then the set $\Gamma'(f, x_*)$ of root classes is finite and every root class is compact.*

Proof. Since $\Gamma(f, x_*)$ is closed in Y from Lemma 3.2 and Y is compact by hypothesis, $\Gamma(f, x_*)$ is compact. Then follows our theorem from Lemma 3.3 (cf. the proof of Corollary II2.4). □

Example 3.1. If A is a non–empty arcwise connected subset of Y and all points of A are roots of $f(y)=x_*$, then all points of A belong to the same root class.

If $f:Y\to x_0$ is a constant mapping, then according as $x_0=x_*$ or $\neq x_*$, $\Gamma'(f, x_*)$ consists of the arcwise connected components of Y or $\Gamma'(f, x_*)=\Gamma(f, x_*)=\emptyset$ *respectively.*

Example 3.2. Let $Y=X=\mathrm{R}^1$, a topological group with addition as the group operation,

$$f:\mathrm{R}^1\to\mathrm{R}^1,\ x\mapsto f(x)=x^2,$$

and $x_*=1$. Then $\Gamma(f,1)=\{-1,1\}$ is the set of all soots of (1) as well as the set of fixed points of the self-mapping $g(x)=x^2+x-1$ in (3). The two roots are of the same root class . If $x_*=-1$, then $\Gamma(f,-1)=\emptyset$.

Example 3.3. Let $Y=X=S^1$, the unit circle in the complex plane (see I § 1) and a topological group with the multiplication of complex numbers as the group operation,

$$f_n:S^1\to S^1,\ z\mapsto f_n(z)=z^n,\ n\neq 0,$$

and $x_*=1$. Then

$$\Gamma(f,\ 1)=\{e^{2\pi ri/|n|},\ r=1,2,\cdots,|n|\},$$

and each root constitutes a root class by itself. The $g(z)$ in (3) now is z^{n+1}.

4. The correspondence between root classes induced by a homotopy

4.1 Definition. Let $H:f_0\simeq f_1:Y\to X$ be a homotopy between f_0 and f_1, x_* a given point in X, and y_i a root of the equation

$$f_i(y)=x_*,\ i=0,1.$$

If there exists in Y a path c from y_0 to y_1 such that

$$\langle\Delta(H,c)\rangle=\langle e_*\rangle$$

(see A § 3 for definition and properties of the diagonal path $\Delta(H,c)$), then y_0 and y_1 are said to be *in correspondence under* H (cf. Definition II 3.1 and Theorem II 4.7) and denoted by y_0Hy_1.

It is easy to see that $y_0Hy_1 \Leftrightarrow y_1H^{-1}y_0$.

Remark 1. If the homotopy H in Definition 4.1 is the constant homotopy $H(\cdot,t)\equiv f$ $(f_i\equiv f,\ i=0.1)$, then Definition 4.1 becomes Definition 3.1. In other words, when H is the constant homotopy, the two roots y_0 and y_1 of the equation $f(y)=x_*$ are in correspondence according to Definition 4.1 (y_0Hy_1 and also y_1Hy_0) if and only if y_0 and y_1 are of the same root class.

Next, let the homotopy H in Definition 4.1 be a closed homotopy but not constant homotopy, and y_0 and y_1 be two roots of the same equation $f(y)=x_*$. Then y_0Hy_1 implies neither that y_0 and y_1 are of the same root class, nor y_1Hy_0.

Remark 2. Even if there exists a homotopy H: $f_0\simeq f_1:Y\to X$, it may happen that $f_0(y)=x_*$ has root while $f_1(y)=x_*$ has not, or that $f_i(y)=x_*$ has roots y_i, $i=0,1$, but not y_0Hy_1.

These happen in the following two examples.

Example 4.1. The homotopy $H(x,t)=x^2+2t$, $0\leqslant t\leqslant 1$, is defined from $f_0(x)=x^2$ to $f_1(x)=x^2+2$, i.e., H: $f_0\simeq f_1:\mathrm{R}^1\to\mathrm{R}^1$. $\Gamma(f_0,1)=\{-1,1\}$, while $\Gamma(f_1,1)=\emptyset$. Hence the root 1 or -1 of the equation $f_0(x)=1$ or $x^2=1$ corresponds to no root of the equation $f_1(x)=1$ or $x^2+1=0$.

Example 4.2. The homotopy $H(z,t)=z^2e^{-\pi ti}$, $0\leqslant t\leqslant 1$, is defined from $f_0(z)=z^2$ to $f_1(z)=z^2e^{-\pi i}$, i.e., $H: f_0\simeq f_1: S^1\to S^1$, where S^1 is the unit circle in the complex plane. Obviously

$$\Gamma(f_0,1)=\{1,-1\},\ \Gamma(f_1,1)=\left\{e^{\frac{\pi i}{2}}, e^{\frac{3\pi i}{2}}\right\}.$$

If one takes the path $c(t)=e^{\frac{\pi ti}{2}}$ in S^1 from root 1 in $\Gamma(f_0,1)$ to the root $e^{\frac{\pi i}{2}}$ in $\Gamma(f_1,1)$, then $\Delta(H,c)(t)$ is the point path $e_1: I\to\{1\}$, and thus the root 1 in $\Gamma(f_0,1)$ corresponds to the root $e^{\frac{\pi i}{2}}$ in $\Gamma(f_1,1)$ under this H.

Next, let c be any path in S^1 from the root 1 in $\Gamma(f_0,1)$ to the other root $e^{\frac{3\pi i}{2}}$ in $\Gamma(f_1,1)$. Then there exists certain integer K such that $c\mathrel{\hat{\simeq}} c_k$, where

$$c_k(t)=e^{(2k+\frac{3}{2})\pi ti},\ 0\leq t\leq 1.$$

But now $\Delta(H,c_k)(t)=e^{(4k+2)\pi ti}$. Thus from Lemma A3.6 and Proposition I 5.6, we have

$$\langle\Delta(H,c)\rangle=\langle\Delta(H,c_k)\rangle\neq\langle e_*\rangle;$$

i.e., the root 1 in $\Gamma(f_0,1)$ does not correspond to the root $e^{\frac{3\pi i}{2}}$ under this H.

It is easy to see that $\Gamma'(f_i,1)$ has two elements, $i=0,1$, and that $\Gamma'(f_0,1)$ and $\Gamma'(f_1,1)$ are in one-to-one correspondence under this H.

4.2 Theorem. *Let there be given a homotopy $H: f_0\simeq f_1: Y\to X$, a point $x_*\in X$, roots $y_i\in\Gamma(f_i,x_*)$, $i=0,1$, and finally y_0Hy_1. Let the root class in $\Gamma'(f_i,x_*)$ containing y_i be denoted by $\mathcal{R}_i$. Then*

(i) $y_0'\in\mathcal{R}_0\Leftrightarrow y_0'Hy_1$,
(ii) $y_1'\in\mathcal{R}_1\Leftrightarrow y_0Hy_1'$.

In other words, the relation y_0Hy_1 induces a correspondence from $\mathcal{R}_0$ to $\mathcal{R}_1$ under H, which is denoted by

$$\mathcal{R}_0H\mathcal{R}_1.$$

It is easy to see that $\mathcal{R}_0H\mathcal{R}_1\Leftrightarrow\mathcal{R}_1H^{-1}\mathcal{R}_0$.

Proof. "$\Rightarrow$" in (i). The hypothesis that $y_0,y_0'\in\mathcal{R}_0\Rightarrow$ there exists in Y a path c_0 from y_0' to y_0 such that $\langle\Delta(f_0,c_0)\rangle=\langle e_*\rangle$, where f_0 is regarded as the constant homotopy. $y_0Hy_1\Rightarrow$ there exists in Y a path c from y_0 to y_1 such that $\langle\Delta(H,c)\rangle=\langle e_*\rangle$. But

$$H\mathrel{\hat{\simeq}} f_0H.$$

Hence, from Lemmas A3.6 and A3.4 (ii), there exists

$$\begin{aligned}\langle\Delta(H,c_0c)\rangle&=\langle\Delta(f_0H,c_0c)\rangle\\&=\langle\Delta(f_0,c_0)\rangle\langle\Delta(H,c)\rangle=\langle e_*\rangle.\end{aligned}$$

"$\Leftarrow$" in (i). The hypothesis that y_0Hy_1 and $y_0'Hy_1$ or $y_1H^{-1}y_0' \Rightarrow$ there exists a path c in Y from y_0 to y_1 and a path c' in Y from y_1 to y_0' such that

$$\langle\Delta(H,c)\rangle = \langle\Delta(H^{-1},c')\rangle = \langle e_*\rangle.$$

From Lemma A3.6 again,

$$\begin{aligned}\langle\Delta(f_0,cc')\rangle &= \langle\Delta(HH^{-1},cc')\rangle\\ &= \langle\Delta(H,c)\rangle\langle\Delta(H^{-1},c')\rangle = \langle e_*\rangle.\end{aligned}$$

(ii) follows immediately from $y_0Hy_1' \Rightarrow y_1'H^{-1}y_0$ and (i) above. □

4.3 Definition. Let $f: Y \to X$ be a mapping and $H: f \simeq H(\cdot,1): Y \to X$ a homotopy. If a root class $\mathcal{R}\,(\in \Gamma'(f, x_*))$ of the equation

$$f(y) = x_* \tag{1}$$

corresponds to a root class $\in \Gamma'(H(\cdot,1),x_*)$ under any such H, then the root class $\mathcal{R}$ is called an *essential root class*. Denote *the set of essential root classes* of (1) by $\Gamma^*(f, x_*)$. The number $\#\,\Gamma^*(f, x_*)$ of elements in $\Gamma^*(f, x_*)$ is called the *Nielsen number* of the equation (1), and is denoted by $N(f, x_*)$.

4.4 Theorem. *If $\mathcal{R} \in \Gamma^*(f, x_*)$, then under any homotopy H with f as the starting point, $\mathcal{R}$ corresponds to one and only one essential root class $\mathcal{R}_1 \in \Gamma^*(H(\cdot,1),x_*)$. Hence under the homotopies H and H^{-1} (Definition 4.1), the correspondences between roots y_0Hy_1 and $y_1H^{-1}y_0$ induce a pair of mutually inverse one-to-one correspondences between the sets of essential root classes $\Gamma^*(f, x_*)$ and $\Gamma^*(H(\cdot,1),x_*)$. Thus*

$$N(f, x_*) = N(H(\cdot,1),x_*),$$

i.e., the Nielsen number $N(f, x_*)$ of the Equation (1) is a homotopy invariant.

Proof. From the hypothesis, $\mathcal{R}$ is essential. Then from Definition 4.3, there is a root class $\mathcal{R}_1 \in \Gamma'(H(\cdot,1),x_*)$ such that $\mathcal{R}H\mathcal{R}_1$. From Theorem 4.2, this $\mathcal{R}_1$ is unique. We still need to prove now that $\mathcal{R}_1$ is also essential, i.e., that, if H' is a homotopy with $H(\cdot,1)$ as the starting point, there exists a root class $\mathcal{R}_2 \in \Gamma'(H'(\cdot,1),x_*)$ $\mathcal{R}_1H'\mathcal{R}_2$. Now we are going to determine $\mathcal{R}_2$ as follows. Since $\mathcal{R}$ is essential, there exists a root class $\mathcal{R}_2 \in \Gamma'(H'(\cdot,1),x_*)$ satisfying $\mathcal{R}(HH')\mathcal{R}_2$; then from $\mathcal{R}_1H^{-1}\mathcal{R}$ and $\mathcal{R}(HH')\mathcal{R}_2$, it follows $\mathcal{R}_1H^{-1}(HH')\mathcal{R}_2 \Rightarrow \mathcal{R}_1H'\mathcal{R}_2$. This means that $\mathcal{R}_1$ is also essential. □

5. Another subgroup $S(X,x_*)$ of the fundamental group $\pi_1(X,x_*)$

In this section, we consider only one topological space X without reference to another space Y and mapping $f: Y \to X$.

5.1 Definition. Let X be a topological space, x_* a given point in X, and $H: X \times I \to X$ a homotopy from $H(\cdot,0)$ to $H(\cdot,1)$ such that $H(\cdot,0)$ and $H(\cdot,1)$ are self homeomorphisms of X with x_* as a fixed point. Hence the diagonal path $\Delta(H,x_*)$ (x_* denoting the point path) is a loop at x_*. For all such homotopies as H, the totality of all the homotopy classes of such loops with both endpoints fixed at x_*, is denoted by $S(X,x_*)$.

5.2 Theorem. *$S(X,x_*)$ is a subgroup of $\pi_1(X,x_*)$. Obviously, the Jiang group (Definition* III4.1) *$J(id,x_*) \subseteq S(X,x_*)$.*

We shall prove first the following lemma.

5.3 Lemma. *If $\alpha \in S(X,x_*)$, then there exists a homotopy $H: X \times I \to X$ such that $H(\cdot,0)$ is a self homeomorphism with x_* as a fixed point, and $H(\cdot,1)$ is the identity mapping id_X of X, and finally $\langle\Delta(H,x_*)\rangle = \alpha$.*

Proof. Since $\alpha \in S(X,x_*)$, there exists a homotopy $H': X \times I \to X$ such that both $H'(\cdot,0)$ and $H'(\cdot,1)$ are self homeomorphisms with x_* as a fixed point and $\langle\Delta(H',x_*)\rangle = \alpha$. Denote the inverse of the homeomorphism $H'(\cdot,1)$ by g, $H'(\cdot,1) \circ g = id_X$. Define $H: X \times I \to X$ by setting $H(x,t) = H'(g(x),t)$. Then $g(x_*) = x_* \Rightarrow \langle\Delta(H,x_*)\rangle = \langle\Delta(H',x_*)\rangle = \alpha$. This H is the homotopy desired. □

Proof of Theorem 5.2. Let $\alpha_1, \alpha_2 \in S(X,x_*)$. We will prove $\alpha_1\alpha_2^{-1} \in S(X,x_*)$. From Lemma 5.3, there exists homotopies $H_i: X \times I \to X$, $i=1,2$, such that $H_i(\cdot,0)$ are self homeomorphisms of X with x_* as a fixed point, $H_i(\cdot,1) = id_X$ and $\langle\Delta(H_i,x_*)\rangle = \alpha_i$. Set $H = H_1 H_2$. From Lemma A3.4(ii), we have

$$\begin{aligned}\alpha_1\alpha_2^{-1} &= \langle\Delta(H_1,x_*)\rangle\langle\Delta(H_2,x_*)\rangle^{-1} = \langle\Delta(H_1H_2^{-1},x_*)\rangle \\ &= \langle\Delta(H,x_*)\rangle \in S(X,x_*).\end{aligned}$$ □

5.4 Theorem. *If X is a manifold (a connected Hausdorff space each point of which has a neighborhood homeomorphic to the n-dimensional Euclidian space, $n>0$), then*

$$S(X,x_*) = \pi_1(X,x_*).$$

Proof [1]. We are going to show that, for any given $\langle c\rangle \in \pi_1(X, x_*)$, there exists a homotopy $H:X\times I\to X$ such that $H(\cdot,0)$ and $H(\cdot,1)$ are self homeomorphisms of X with x_* as a fixed point and $\langle\Delta(H,x_*)\rangle=\langle c\rangle$.

For any given positive integer m, we may divide the given c into m parts, and turn the i-th part into a path $c_i:I\to X$, $i=1,2,\cdots,m$, so that $c \simeq c_1c_2\cdots c_m$. Since X is an n-dimensional manifold, for appropriate choice of m, there exist open sets $U_i\subseteq X$ and homeomorphisms h_i from the closure $\bar{U}_i$ to the same unity spherical ball B^n (with boundary) in $\mathbb{R}^n$, such that $c_i(I)\subseteq U_i$ and $h_i(c_i(1))$ is the center of B^n, the origin O in $\mathbb{R}^n$, $i=1,2,\cdots,m$. For each given i, define a homotopy $H_i:X\times I\to X$ by

$$H_i(x,t)=\begin{cases}h_i^{-1}(h_i^{-1}(x)+(1-|h_i(x)|)h_i(c_i(t))), & \text{when } x\in \bar{U}_i;\\ x, & \text{when } x\notin \cup U_i.\end{cases}$$

Remark. A homotopy $g:B^n\times I\to B^n$ with every boundary point of B^n fixed under $g(\cdot,t)$.

For an interior point D of B^n, we define a homeomorphism $g:B^n\to B^n$ with every boundary point fixed by means of the following

$$g:B^n\to B^n,\ y\mapsto g(y)=y+(1-|y|)D,$$

as shown in Figure 2. Geometrically speaking, any interior point y of B^n,

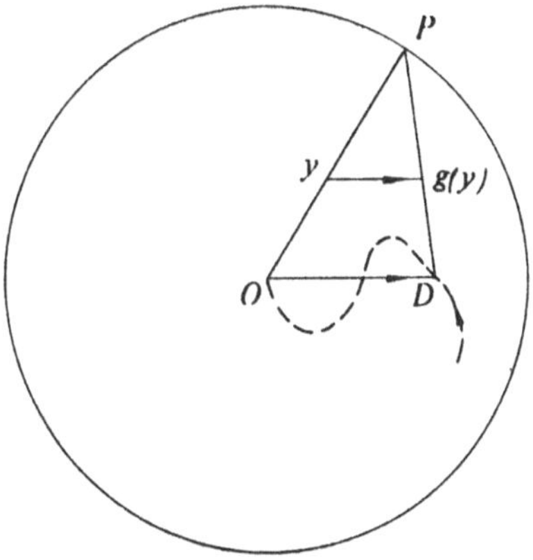

Figure 2

when $\neq O$, determines with O a boundary point P of B^n, and, when not on the straight line OD, determines with O and D a 2-plane, and thus finally determines the point $g(y)$ such that $yg(y)\parallel OD$. On removing the cases $y=0$ and y

---

1) The proof in [17] is based on certain theorems in the theory of fibre bundles. As the reader may not be familiar with this theory, we use here a more elementary proof due to Jiang Boju, which is a modification of the proof in the preprint of [17].

on OD by the principle of continuity, we see g is a self homeomorphism as required.

Furthermore, suppose there is a path $d:I\to$ interior of B^n with $d(1)=0$. As the point $D=d(t)$ determines the g above, the path d gives rise to a homotopy $G:B^n\times I\to B^n$, $(y,t)\mapsto G(y,t)=y+(1-|y|)d(t)$. It is important to note that $G(\cdot,t)$ should always be a homeomorphism for any given value of t, $G(\cdot,1)=id$, and finally

$$\Delta(G(y),\ d(1))(t)=G(d(1),t)=d(t).$$

On identifying respectively $d(t)$, y and $g(y)$ in Figure 2 with $h_i(c_i(t))$, $h_i(x)$ and $h_i(x)+(1-|h_i(x)|)h_i(c_i(t))$ in the formula of $H_i(x,t)$, we find at once that $H_i(\cdot,t)$ is always a homeomorphism for any given t, $H_i(\cdot,1)=id_X$, and $\Delta(H_i,c_i(1))(t)=c_i(t)$, $i=1,2,\cdots,m$.

Next, set

$$H'_m=H_m,\text{ and } H'_i(x,t)=H_i(H'_{i+1}(x,0),t), i=1,2,\cdots,m-1,$$

in the order from $i=m-1,\ m-2,\cdots,1$. We have

$$H'_i(x,1)=H_i(H'_{i+1}(x,0),1)=H'_{i+1}(x,0).$$

On the other hand, from $\Delta(H'_m,x_*)=\Delta(H_m,x_*)=c_m$, we have

$$\begin{aligned}\Delta(H'_{m-1},x_*)(t)&=H_{m-1}(H'_m(x_*,0),t)=H_{m-1}(H_m(x_*,0),t)\\&=H_{m-1}(c_m(0),t)=H_{m-1}(c_{m-1}(1),t)=c_{m-1}(t),\end{aligned}$$

i.e.

$$\Delta(H'_{m-1},x_*)=c_{m-1},$$

and for general $i<m$

$$\Delta(H'_i,x_*)=c_i.$$

Finally, define the homotopy $H:X\times I\to X$ by $H=H'_1H'_2\cdots H'_m$. Then from Lemma A3.4(ii), we deduce

$$\begin{aligned}\langle\Delta(H,x_*)\rangle&=\langle\Delta(H'_1H'_2\cdots H'_m,x_*)\rangle\\&=\langle\Delta(H'_1,x_*)\rangle\langle\Delta(H'_2,x_*)\rangle\cdots\langle\Delta(H'_m,x_*)\rangle\\&=\langle c_1\rangle\langle c_2\rangle\cdots\langle c_m\rangle=\langle c\rangle.\end{aligned}$$

□

6. Reidemeister number of equation

Let the equation

$$f(y)=x_*$$

be that in Definition 3.1. Moreover Let $y_*\in\Gamma(f,x_*)$. This implies

$\Gamma(f, x_*) \neq \emptyset$, and hence also $\Gamma'(f, x_*) \neq \emptyset$. Then f induces a homomorphism

$$f_\pi : \pi_1(Y, y_*) \to \pi_1(X, x_*).$$

Denote *the set of right cosets of* $\operatorname{Im} f_\pi$ in $\pi_1(X, x_*)$ by Cosets $\operatorname{Im} f_\pi$, and call the number of its elements the *Reidemeister number* $R(f, x_*)$ of the equation:

$$R(f, x_*) = \# \text{ Cosets } \operatorname{Im} f_\pi.$$

Remark 1. Let $y_* \in \mathscr{R} \in \Gamma'(f, x_*)$. Then $\operatorname{Im} f_\pi$ and Cosets $\operatorname{Im} f_\pi$ are obviously independent of the choice of y_* in $\mathscr{R}$, and $R(f, x_*)$ is independent of the choice of $\mathscr{R}$ in $\Gamma'(f, x_*)$.

The following theorem relates the root classes to the Reidemeister number $R(f, x_*)$.

6.1 Theorem. *Let $\mathscr{R}_* \in \Gamma'(f, x_*)$ be a given root class. For any root class $\mathscr{R}_1 \in \Gamma'(f, x_*)$, take any $y_* \in \mathscr{R}_*$, any $y_1 \in \mathscr{R}_1$ and any path c in Y from y_* to y_1, and define the correspondence*

$$\varphi : \Gamma'(f, x_*) \to \text{Cosets } \operatorname{Im} f_\pi, \quad \mathscr{R}_1 \mapsto \varphi(\mathscr{R}_1) = (\operatorname{Im} f_\pi) \cdot \langle f \circ c \rangle.$$

Then

(i) *φ is single-valued, i.e., is independent of the choice of c in Y, the choice of y_* in $\mathscr{R}_*$ and the choice of y_1 in $\mathscr{R}_1$;*

(ii) *φ is injective; and thus*

$$\# \Gamma'(f, x_*) \leqslant R(f, x_*).$$

Proof. (i) Let c' be another path in Y from y_* to y_1. Then $c' = bc$, where b is a loop in Y at y_*. Then

$$\langle f \circ c' \rangle = \langle f \circ b \rangle \langle f \circ c \rangle \in \operatorname{Im} f_\pi \cdot \langle f \circ c \rangle, \text{ i.e.,}$$

$$\operatorname{Im} f_\pi \cdot \langle f \circ c' \rangle = \operatorname{Im} f_\pi \cdot \langle f \circ c \rangle.$$

Let y_1' be another root $\in \mathscr{R}_1$. As y_1 and $y_1' \in$ the same $\mathscr{R}_1$, there exists in Y a path b' from y_1 to y_1' such that $\langle f \circ b' \rangle = \langle e_* \rangle$. Take the path $c' = cb'$ from y_* to y_1'. We find $\langle f \circ c' \rangle = \langle f \circ c \rangle$, and thus again

$$\operatorname{Im} f_\pi \cdot \langle f \circ c' \rangle = \operatorname{Im} f_\pi \cdot \langle f \circ c \rangle.$$

Similarly φ can be proved to be independent of the choice of y_* in $\mathscr{R}_*$

(ii) In order to show that φ is injective, let $\mathscr{R}_1, \mathscr{R}_1' \in \Gamma'(f, x_*)$ and $\varphi(\mathscr{R}_1) = \varphi(\mathscr{R}_1')$. Take any $y_1 \in \mathscr{R}_1$, any $y_1' \in \mathscr{R}_1'$, and any paths c_1, c_1' from y_* to y_1, y_1' respectively. Then

$$\varphi(\mathcal{R}_1)=\varphi(\mathcal{R}_1')\Rightarrow \mathrm{Im} f_\pi \cdot \langle f\circ c_1\rangle = \mathrm{Im} f_\pi \cdot \langle f\circ c_1'\rangle.$$

This $\Rightarrow \langle f\circ c_1'\rangle \in \mathrm{Im} f_\pi \cdot \langle f\circ c_1'\rangle \Rightarrow$ there exists in Y a loop b at y_* such that $\langle f\circ c_1'\rangle = \langle f\circ b\rangle\langle f\circ c_1\rangle \Rightarrow \langle f\circ c_1'^{-1} b c_1\rangle = \langle e_*\rangle$. But this shows $\mathcal{R}_1=\mathcal{R}_1'$ by Definition 3.1 □

6.2 Lemma. *Let $\mathcal{R}_*$ be an essential root class, i.e., $\mathcal{R}_*\in\Gamma^*(f, x_*)$. If $\alpha\in S(X,x_*)$, then there exists an essential root class $\mathcal{R}_1\in\Gamma^*(f, x_*)$ such that*

$$\varphi(\mathcal{R}_1)=(\mathrm{Im} f_\pi)\cdot\alpha.$$

Proof. 1) Since $\alpha\in S(X,x_*)$, from Lemma 5.3 there exists a homotopy H of self-mappings of X such that $H(\cdot,0)$ is a homeomorphism and $H(\cdot,1)=id_X$, both keeping the point x_* fixed, and $\langle\Delta(H,x_*)\rangle=\alpha$.

2) Set $F(y,t)=H(f(y),t)$, $y\in Y$, $t\in I$, and thus we get a homotopy $F:H(\cdot,0)\circ f\simeq f$. Take a root $y_*\in\mathcal{R}_*\in\Gamma^*(f, x_*)$. From Theorem 4.4, there exists a unique essential root class $\mathcal{R}_1\in\Gamma^*(H(\cdot,0)\circ f,x_*)$ of the equation $H(\cdot,0)\circ f(y)=x_*$ such that $\mathcal{R}_1 F\,\mathcal{R}_*$. Take a root $y_1\in\mathcal{R}_1$. Then $y_1 F y_*$.

3) Since x_* is a fixed point of the self homeomorphism $H(\cdot,0)$ of X, the two equations $f(y)=x_*$ and $H(\cdot,0)\circ f(y)=x_*$ have the same set of roots; and moreover the root class of equation $f(y)=x_*$ which contains y_1 is just $\mathcal{R}_1$ mentioned in 1). It remains to be proven that $\mathcal{R}_1$ is the essential root class whose existence is asserted in the conclusion of our theorem. This proof consists of the following two steps: 4) $\mathcal{R}_1$ is an essential root class of the equation $f(y)=x_*$, *i.e.* $\mathcal{R}_1\in\Gamma^*(f, x_*)$; and 5) $\varphi(\mathcal{R}_1)=\mathrm{Im} f_\pi\cdot\alpha$.

4) In order to show $\mathcal{R}_1\in\Gamma^*(f, x_*)$, from Definition 4.3 for any homotopy $G: f\simeq G(\cdot,1)$ we must determine a root $y_2\in\Gamma(G(\cdot,1), x_*)$ such that $y_1 G\, y_2$. Now define a homotopy G' by $G'(\cdot,t)=H(\cdot,1)\circ G(\cdot,t)$ which starts from $H(\cdot,0)\circ f$. Since from 2), $y_1\in\mathcal{R}_1\in\Gamma^*(H(\cdot,0)\circ f,x_*)$, then from Definition 4.3 there exists a root y_2 of the equation $G'(y,1)=x_*$ such that $y_1 G' y_2$, i.e., there exists in Y a path c from y_1 to y_2 such that $\langle\Delta(G',c)\rangle=\langle e_*\rangle$. Since

$$\begin{aligned}\langle\Delta(G,c)\rangle&=\langle\Delta(H(\cdot,0)^{-1}\circ G',c)\rangle=(H(\cdot,0)^{-1})_\pi\cdot\langle\Delta(G',c)\rangle\\&=(H(\cdot,0)^{-1})_\pi\cdot\langle e_*\rangle=\langle e_*\rangle,\end{aligned}$$

we have the desired $y_1\ G\ y_2$.

5) The fact $y_1 F y_*$ in 2) $\Rightarrow$ there exists in Y a path D from y_1 to y_* such that $\langle\Delta(F,D)\rangle=\langle e_*\rangle\cdot\alpha=\langle\Delta(H,x_*)\rangle$ is the conclusion in 1). Now, $F(y,t)=H(f(y),t)$ and $f(y_*)=x_*\Rightarrow\langle\Delta(H,x_*)\rangle=\langle\Delta(F,y_*)\rangle$.

Moreover, from Lemmas A3.6 and A3.4 (ii), we have

$$\begin{aligned}\langle\Delta(F,y_*)\rangle&=\langle\Delta(Ff,DD^{-1})\rangle=\langle\Delta(F,D)\rangle\langle\Delta(f,D^{-1})\rangle\\&=\langle e_*\rangle\langle f\circ D^{-1}\rangle=\langle f\circ D^{-1}\rangle.\end{aligned}$$

Thus
$$\alpha=\langle f\circ D^{-1}\rangle,$$

where D^{-1} is a path in Y from y_* to y_1; i.e., from the definition of φ,

$$\varphi(\mathcal{R}_1)=\mathrm{Im}f_\pi\cdot\alpha. \qquad \square$$

6.3 Definition. Let $f:(Y,y_*)\to(X,x_*)$ be a mapping, and hence Cosets $\mathrm{Im}f_\pi$ is defined. When $\alpha\in S(X,x_*)$, we shall say that the element $\mathrm{Im}f_\pi\cdot\alpha$ of Cosets $\mathrm{Im}f_\pi$ has a *representative* in $S(X,x_*)$. As every element of $S(X,x_*)$ is a representative of a certain right coset, two elements of $S(X,x_*)$ will be said to belong to the same *representative class* when they are representatives of the same right coset. We denote *the number of the representative classes in* $\mathbf{S}(X,x_*)$ *by* $\mathbf{s}(f,y_*)$.

Remark 2. By Definition 5.1, $S(X,x_*)$ has nothing to do with f. But as $\mathrm{Im}\,f_\pi$ depends on both f and the existence of a root y_* of $f(y)=x_*$, so does the number $s(f,y_*)$. Compare Remark 1.

On the basis of Definitions 6.3 and 4.3, there follows from Theorem 6.1 and Lemma 6.2 the following.

6.4 Theorem. *If y_* belongs to an essential root class of the equation $f(y)=x_*$, then*

$$s(f,\ y_*)\leqslant N(f,\ x_*). \qquad \square$$

6.5 Corollary. *If $S(X,x_*)=\pi_1(X,x_*)$, and the equation $f(y)=x_*$ has at least one essential root class, then every root class of the equation is essential and $N(f,\ x_*)=R(f,\ x_*)$.*

Proof. From Theorems 6.4 and 6.1, there follows

$$s(f,\ y_*)\leqslant N(f,\ x_*)\leqslant\#\Gamma'(f,\ x_*)\leqslant R(f,\ x_*).$$

On the other hand, from $S(X,\ x_*)=\pi_1(X,\ x_*)$ and definitions of $s(f,\ y_*)$ and $R(f,\ x_*)$, we have $s(f,\ y_*)=R(f,\ x_*)$. Thus

$$N(f,\ x_*)=\#\Gamma'(f,\ x_*)=R(f,\ x_*). \qquad \square$$

From Theorem 5.4 and Corollary 6.5, we have immediately the following

6.6 Corollary. *If Y is a manifold and the equation $f(y)=x_*$ has at least one essential root class, then every root class is essential and $N(f,\ x_*)=R(f,\ x_*)$.* $\square$

7. Index of root class. Evaluation of $N(f, x_*)$ in the case of maximal $S(X,x_*)$

Theorems 6.1 and 6.4 give respectively only upper and lower bounds of the Nielsen number $N(f, x_*)$ of the equation $f(y)=x_*$. In the present section, on making further use of the subgroup $S(X,x_*)$ of $\pi_1(X,x_*)$ (just as making use of the Jiang group in Chapter III), we shall sketch without detailed derivation the results on the evaluation of $N(f, x_*)$ in the case of the maximal $S(X,x_*)$. (Theorem 7.1 is similar to Theorem III5.1 and Corollary 7.3 which has no counterpart in Chapter III.)

The concept of index is again the key to the derivation of these results. As the spaces under consideration are no longer polyhedra but arcwise connected topological spaces, we have to turn to the singular homology theory instead of the simplicial one we have used so far. In order to make the discussion not too long, we are only going to sketch the results without details and proofs.

Now Let the topological space Y be compact, normal, arcwise connected and locally arcwise connected, and X be arcwise connected and locally arcwise connected Hausdorff space. Then any root class $\mathcal{R}$ of the equation $f(y)=x_*$ has a neighborhood U open in Y, whose closure $\overline{U}$ contains no other root $\in \Gamma(f, x_*)$. Consider first the mappings

$$Y \xrightarrow{i} (Y, Y-\mathcal{R}) \xrightarrow{e} (\overline{U}, \overline{U}-\mathcal{R}) \xrightarrow{f'} (X, X-x_*),$$

where i and e are inclusions, e is an excision in addition, and f' is the restriction of f on $\overline{U}$. They induce the homomorphism

$$f'_* \circ e_*^{-1} \circ i_* : H_*(Y) \to H_*(X, X-x_*),$$

where H_* denotes the sum of the singular homology groups of all dimensions. Next, consider the mappings

$$Y \xrightarrow{f} X \xrightarrow{j} (X, X-x_*),$$

where j is again the inclusion. They induce the homomorphism

$$j_* \circ f_* : H_*(Y) \longrightarrow H_*(X, X-x_*).$$

Now call the homomorphism

$$v(f, x_*; \mathcal{R}) = f'_* \circ e_*^{-1} \circ i_*$$

and

$$v(f, x_*) = j_* \circ f_*$$

the *index* of the root class $\mathscr{R}$ and the *index* of $\Gamma(f, x_*)$ respectively. Note that the index is now no longer an integer but a homomorphism.

It can be shown that the index here has also the ordinary properties of the ordinary local index theory, in particular, the following properties of additivity and homotopy invariance. The property of additivity states

$$v(f,x_*) = \sum_{R \in \Gamma'(f,x_*)} v(f,x_*;\mathscr{R}).$$

To state the property of homotopy invariance, consider any homotopy $H: f \simeq f'$. Then the statement is as follows. If the root class $\mathscr{R}$ of f corresponds under H to the root class $\mathscr{R}'$ of f', then

$$v(f, x_*;\mathscr{R}) = v(f', x_*;\mathscr{R}');$$

if $\mathscr{R}$ corresponds to no root class of f' under H, then

$$v(f, x_*;\mathscr{R}) = 0.$$

After this sketch of the definition of index, we can now state the results as promised: Theorem 7.1 for general Y and X and Corollary 7.3 for manifolds Y and X.

7.1 Theorem. *If $j_* \circ f_* \neq 0$, then $N(f, x_*) > 0$. If, in addition, $S(X, x_*) = \pi_1(X, x_*)$ and j_* is epimorphic, then every root class in $\Gamma'(f, x_*)$ is essential, $N(f, x_*) = R(f, x_*)$, and the indices of all root classes equal, i.e.,*

$$v(f, x_*) = R(f, x_*) \cdot v(f, x_*;\mathscr{R}),$$

where $\mathscr{R}$ is any given root class. □

7.2 Definition. Let both X and Y be n-dimensional oriented closed manifolds, and μ and ν the respective generators of their n-dimensional homology groups with integral coefficients. Then

$$f_{*n}(\nu) = (\deg f)\mu,$$

and

$$j_{*n}: H_n(X) \to H_n(X, X - x_*)$$

is isomorphic. If for a root class $\mathscr{R} \in \Gamma'(f, x_*)$

$$v(f, x_*;\mathscr{R})(\nu) = m(\mathscr{R})\mu,$$

then $m(\mathscr{R})$ is called the *multiplicity* of $\mathscr{R}$.

7.3 Corollary. *If both X and Y are oriented closed manifolds of the same dimension and $\deg f \neq 0$, then every root class is essential,*

$$N(f, x_*) = R(f, x_*),$$

all root classes have the same multiplicity, and the sum of the multiplicities of all the root classes equals $\deg f$. □

Appendix A

HOMOTOPY AND FUNDAMENTAL GROUP [1)]

1. Homotopy

The unit closed interval $[0,1]=\{t: 0\leqslant t\leqslant 1\}$is denoted by I.

1.1 Definition. Let X and Y be topological spaces and f_0, f_1:$X \to Y$ be mappings. If there exists a mapping

$$F: X\times I\to Y$$

such that $F(x,i)=f_i(x),\forall x\in X$ or $F(\cdot,i)=f_i$, $i=0,1$, then f_0 and f_1 are said to be *homotopic* and are denoted by

$$F: f_0\simeq f_1, \text{ or } F: f_0\simeq f_1: X\to Y.$$

The mapping F is called a *homotopy from f_0 to f_1*. If $f_0=f_1=f$ and $F(\cdot,t)\equiv f$, then F is called the *constant homotopy of f.*

We often denote $F(\cdot,t)$by f_t,and represent the homotopy also in the form f_t: $f_0\simeq f_1$.

1.2 Theorem. *Let $M(X, Y)$ be the set of all mappings from X to Y. The homotopy relation is an equivalence relation in $M(X, Y)$.*

1.3 Definition. The subset of mappings homotopic to a given mapping f: $X\to Y$ is called the *mapping class* of f, and denoted by $\langle f\rangle$.

1.4 Corollary. *$M(X, Y)$ is separated into pairwise disjoint mapping classes.*

---

1) Added to the English version——In the 1979 Chinese edition of this book, most of the theorems and statements in the Appendices A and B are with proofs, based on some current texts in English or in Chinese, such as [8]—[12],in particular [10] and [12]. In order to make the English version shorter, we omit all such proofs and all easier examples. We recommend [3] to take the place of [8] (in Chinese).

1.5 Theorem. *If X, Y and Z are topological spaces, and if there are homotopies* $f_0 \simeq f_1: X \to Y$ *and* $g_0 \simeq g_1: Y \to Z$, *then* $g_0 \circ f_0 \simeq g_1 \circ f_1: X \to Z$.

2. Path. Product and inverse. Subpath

Now restrict to mappings: $I \to X$, where $I = \{s: 0 \leqslant s \leqslant 1\}$.

2.1 Definition. Let X be a topological space. A mapping $a: I \to X$ is called a path in X. The points $a(0) = x_0$ and $a(1) = x_1$ are called the *starting point* and *terminal point* respectively, but both are called the *endpoints*. If the endpoints coincide, the path is called a *loop*. If $a(I) = \{x\}$, x being a single point of X, the path is called a *constant path* or *point path*. A point path at point x, or x_0, or x_1 will be denoted by $e_x: I \to x$, or $e_0: I \to x_0$, or $e_1: I \to x_1$ respectively.

2.2 Definition. Let a be a path from x_0 to x_1 in X and b a path from x_1 to x_2 in X. The *product* ab: $I \to X$ of the paths a and b is defined by the following mapping.

$$ab(s) = \begin{cases} a(2s), & 0 \leqslant s \leqslant \frac{1}{2}; \\ b(2s-1), & \frac{1}{2} \leqslant s \leqslant 1. \end{cases}$$

The *inverse* a^{-1}: $I \to X$ of the path a is defined by the following mapping

$$a^{-1}(s) = a(1-s), 0 \leqslant s \leqslant 1.$$

2.3 Definition. Let the path $a: I \to X$ be given. For two constants $r, s \in I$, $r \leqslant s$, the mapping

$$a_r^s: I \to X, t \mapsto a_r^s(t) = a(r + (s-r)t)$$

is called a *subpath* of a. It is the restriction of a on the interval $[r, s]$. When $s \leqslant r$, a_r^s is a subpath of a^{-1}, the restriction of a^{-1} on the interval $[s, r]$.

3. Two types of path classes

3.1 Definition. Let a and b be two paths in X. By Definition 1.1, if there exists a mapping

$$F: I \times I \to X, (s, t) \mapsto F(s, t)$$

such that $F(s, 0)$ and $F(s, 1)$ are $a(s)$ and $b(s)$ respectively, the two paths are homotopic. For paths in particular, we shall call this homotopy a *free homotopy* from a to b, denoted still by $F: a \simeq b$. If a and b have the same endpoints x_0 and x_1, and F satisfies the additional conditions: $F(0, t) \equiv x_0$ and $F(1, t) \equiv x_1$, then the two paths are said to be *homotopic with fixed endpoints* and denoted by $F: a \fallingdotseq b$. All paths with the same two endpoints are separated also into pairwise disjoint path classes, called *path classes with fixed endpoints.*

Let X and Y be topological spaces and $H: f_0 \simeq f_1: Y \to X$ a homotopy from mapping f_0 to mapping f_1, and let $c: I \to Y$ be a path in Y from starting point y_0 to terminal point y_1. Then $H(c(t),t), t \in I$, is a path in X. In order to emphasize the geometrical meaning, we introduce for this path the following notation and terms (see [2] VI D).

3.2 Notation. Let X and Y be topological spaces. For a given homotopy $H: f_0 \simeq f_1: Y \to X$ and a given path $c: I \to Y$, define the path $\Delta(H, c): I \to X$ by

$$\Delta(H, c)(t) = H(c(t), t), 0 \leqslant t \leqslant 1.$$

Δ and $\Delta(H, c)$ are called respectively the *diagonal operation* and the *diagonal path* of H and c (see the figure). The inverse of the diagonal path is denoted by $\Delta^{-1}(H,c)$.

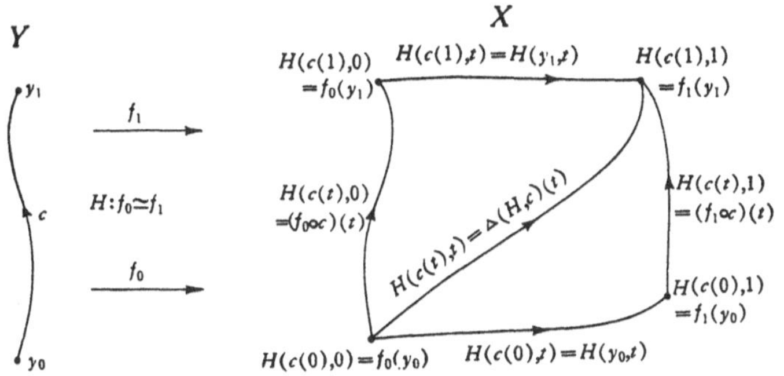

3.3 Definition. Let H and H': $Y \times I \to X$ be homotopies between mappings from Y to X. If $H(y, 1) = H'(y, 0)$, $\forall y \in Y$, then the *product* HH' of the homotopies H and H' is defined by

$$HH'(y, t) = \begin{cases} H(y, 2t), & 0 \leqslant t \leqslant \frac{1}{2}, \\ H'(y, 2t-1), & \frac{1}{2} \leqslant t \leqslant 1, \end{cases}$$

and the *inverse* H^{-1} of the homotopy H by

$$H^{-1}(y, t) = H(y, 1-t), \ 0 \leqslant t \leqslant 1.$$

3.4 Lemma. *The diagonal operation* Δ *has the following two properties*:

(i) *preservation of inverse*: $\Delta(H^{-1}, c^{-1}) = \Delta^{-1}(H, c)$;

(ii) *preservation of product*: *if* $H(y, 1) = H'(y, 0)$ $\forall y \in Y$ *and* $c(1) = c'(0)$, *then* $\Delta(H, c) \cdot \Delta(H', c') = \Delta(HH', cc')$.

3.5 Definition. Let H and H': $Y \times I \to X$ be homotopies between mappings from Y to X. If $H(y, i) = H'(y, i) = f_i(y), i = 0,1$, $\forall y \in Y$, and there exists a mapping

$$\Psi : Y \times I \times I \to X, \ (y, t, s) \mapsto \Psi(y, t, s)$$

such that

$$\Psi(y, t, 0) = H(y, t), \ \Psi(y, t, 1) = H'(y, t), \ \forall y \in Y, \ \forall t \in I,$$
$$\Psi(y, i, s) = f_i(y), \ i = 0,1, \ \forall y \in Y, \ \forall s \in I,$$

the homotopies H and H' are said to be *homotopic with fixed ends* and denoted also by $H \risingdotseq H'$.

3.6 Lemma. *Let the paths* c *and* c' *in* Y *be homotopic with fixed endpoints, and the homotopies* H *and* H' *between mappings from* Y *to* X *be homotopic with fixed ends. Then we have the diagonal paths* $\Delta(H, c) \risingdotseq \Delta(H', c')$.

4. From path classes with fixed endpoints to fundamental group

4.1 Notation. Let a be a path. In this and the succeeding sections, $\langle a \rangle$ always means the path class with fixed endpoints of the path a, and is called for brevity the path class of a.

4.2 Lemma. *If the paths* $a_0 \risingdotseq a_1$, $b_0 \risingdotseq b_1$ *and one of the products* $a_0 b_0$ *and* $a_1 b_1$ *is defined, then the other product is also defined. Furthermore* $a_0 b_0 \risingdotseq a_1 b_1$.

4.3 Lemma. *If the paths* $a \risingdotseq b$, *then* $a^{-1} \risingdotseq b^{-1}$.

4.4 Definition. The *product* of two path classes $\langle a \rangle$ and $\langle b \rangle$ is $\langle a \rangle \langle b \rangle = \langle ab \rangle$ when ab is defined, and the *inverse* of a path class $\langle a \rangle$ is $\langle a \rangle^{-1} = \langle a^{-1} \rangle$.

4.5 Theorem. *Let* a *be a path from* x_0 *to* x_1 *in* X *and* e_i: $I \to x_i$, $i = 0,1$, *the point paths (see Definition 2.1). Then*

(i) $\langle e_0\rangle\langle a\rangle=\langle a\rangle$;
(ii) $\langle a\rangle\langle e_1\rangle=\langle a\rangle$;
(iii) $\langle a\rangle\langle a^{-1}\rangle=\langle e_0\rangle, \langle a^{-1}\rangle\langle a\rangle=\langle e_1\rangle$;
(iv) $(\langle a\rangle\langle b\rangle)\langle c\rangle=\langle a\rangle(\langle b\rangle\langle c\rangle)$, *when* $(ab)c$ *is defined.*

Consider only loops at a specified point of X, we arrive at the following important corollary.

4.6 Corollary. *Let X be a topological space and x_0 a given point of X. The loop classes at x_0 (the path classes with fixed endpoints coincident at x_0) constitute a group under the two operations of forming the product and the inverse. The group is called the fundamental group or the first homotopy group of X at the base point x_0, and is denoted by $\pi_1(X, x_0)$.*

5. Basic properties of $\pi_1(X, x_0)$

5.1 Definition. A topological space is said to be *arcwise connected,* if there exists in the space a path connecting every pair of its points.

5.2 Theorem. *Let X be an arcwise connected topological space, x_0 and x_1 two points of X, and w a path from x_0 to x_1 in X. Then $\pi_1(X, x_0)$ and $\pi_1(X, x_1)$ are isomorphic, denoted by $\pi_1(X, x_0)\approx\pi_1(X, x_1)$, and the isomorphism is given by*

$$w_*: \pi_1(X, x_1)\rightarrow\pi_1(X,x_0), \quad \langle a\rangle \mapsto \langle w\ a\ w^{-1}\rangle\ .$$

5.3 Corollary. *If w_1 and w_2 are two paths from x_0 to x_1 in the arcwise connected space X, then $(w_1)_*\circ(w_2^{-1})_*$ is an inner automorphism of $\pi_1(X, x_0)$.*

5.4 Definition. An arcwise connected topological space is said to be *simply connected,* when its fundamental group is trivial.

5.5 Lemma. *If X is simply connected, then any two paths a_1 and a_2 with the same starting and terminal points are homotopic with fixed endpoints: $a_1 \risingdotseq a_2$.*

5.6 Definition. Let X and Y be arcwise connected topological space, and f: $X\rightarrow Y$ a mapping. If a_1 and a_2 are two homotopic loops at x_0 in X, $a_1\risingdotseq a_2$, then $f\circ a_1\risingdotseq f\circ a_2$. Thus there arises a correspondence

$$f_\pi: \pi_1(X, x_0)\rightarrow\pi_1(Y, f(x_0)),\ \langle a\rangle \mapsto \langle f\circ a\rangle\ .$$

f_π is a homomorphism, called the *homomorphism induced* by f.

5.7 Theorem. *Let X, Y, Z be arcwise connected spaces, and* $x_0 \in X$.

(i) *If* $f: X \to Y$, *and* $g: Y \to Z$ *are mappings, then*

$$(g \circ f)_\pi = g_\pi \circ f_\pi.$$

(ii) *If* $F: f_0 \simeq f_1: X \to Y$ *is a homotopy from mapping* f_0 *to mapping* f_1, *then*

$$(f_0)_\pi = w_* \circ (f_1)_\pi,$$

where w *is a path*

$$w(t) = F(x_0, t)$$

in Y *from* $f_0(x_0)$ *to* $f_1(x_0)$.

5.8 Theorem. *Let* X *be a triangulable and arcwise connected topological space,* $x_0 \in X$, *and* $H_1(X)$ *the* 1*-dimensional homology group of* X *with the group of integers as the coefficient group. Then there exists a natural surjective homomorphism* $\theta: \pi_1(X, x_0) \to H_1(X)$ *such that Ker* θ *is the commutator subgroup. Moreover,*

(i) $\theta \circ w_* = \theta$, *where* w *is a path from* x_0 *to* x_1 *in* X;

(ii) *for a mapping* f *from* X *to a triangulable and arcwise connected space* Y, $\theta \circ f_\pi = f_{1*} \circ \theta$, *where* $f_{1*}: H_1(X) \to H_1(Y)$ *is the homomorphism induced by* f.

Appendix B

COVERING SPACES

A formal or abstract definition of a covering space $\tilde{X}$ of an arcwise connected and locally arcwise connected topological space X is given in Definition 1.2. For a covering space so defined, our first aim is to derive Theorem 2.4 on the lifting of a self-mapping f of X to a self-mapping of $\tilde{X}$, and Theorem 2.5 on the lifting of a homotopy between self-mappings of X to a homotopy of self-mappings of $\tilde{X}$. This is essentially accomplished via Theorems 1.3 (on the lifting of path), Theorem 1.5 (on the lifting of homotopy between paths) and their generalizations —— Lemmas 2.1 and 2.2.

Next, the existence problem of $\tilde{X}$ for a given X arises naturally. On the basis of the properties of the abstract $\tilde{X}$ obtained in §§ 2-3, § 4 deals with the geometrical construction of $\tilde{X}$ in terms of path classes in X when X is locally simply connected (Theorem 4.1).

Finally, for the special case that $\tilde{X}$ is the universal (or simply connected) covering space of X in § 5, the geometrical construction of $\tilde{X}$ leads to the geometrical formulas of the lifting of a self-mapping of $\tilde{X}$ in Theorem 5.1 and of the lifting of a homotopy between self-mappings of X in Theorem 5.4. These two theorems derived for universal $\tilde{X}$ are simpler and more precise than the two corresponding Theorems 2.4 and 2.5 for general coverings, and are just what we need in Chapter III.

Many lemmas, theorems and corollaries in this appendix come from [10], Chapter 5, especially pp. 151—160, and [12], Chapter 3, especially pp. 62—67. Their proofs are omitted.

1. Formal or abstract definition of covering spaces. Two basic theorems on liftings of paths

1.1 Definition. A topological space X is *locally arcwise con-*

nected, if, for any of its point x and any of its open set U containing x, there exists in it an arcwise connected open set V such that $x \in V \subseteq U$.

1.2 Definition. Let both X and $\tilde{X}$ be arcwise connected and locally arcwise connected topological spaces and $p: \tilde{X} \to X$ a mapping onto. The pair $(\tilde{X}, p)$ or simply $\tilde{X}$ is called a *covering space* of X, and the mapping p the *projection,* when for every point x of X, there exists in X an arcwise connected open neighborhood U such that every arcwise connected component of $p^{-1}(U)$ is an open set of $\tilde{X}$ and that p is a homeomorphism from every component to U. Such a neighborhood U is called an *admissible neighborhood* of X.

Let $(\tilde{X}, p)$ be a covering space of X, and U an admissible neighborhood of X. Denote the fibre $p^{-1}(x)$ by $\{\tilde{x}_i\}$, where $x \in U$ and i runs through an index set, and denote the arcwise connected component of $p^{-1}(U)$ containing $\tilde{x}_i$ by $\tilde{U}_i$. $\tilde{U}_i$is an arcwise connected open neighborhood of $\tilde{x}_i$ in $\tilde{X}$. One sees easily that Definition 1.2 implies the following facts.

(i) $p^{-1}(U) = \bigcup_i \tilde{U}_i$, and $\tilde{U}_i \cap \tilde{U}_j = \emptyset$, $i \neq j$.

(ii) As asubspace of $\tilde{X}$, $p^{-1}(x)$ is discrete; i.e., every point $\tilde{x}_i$ is closed as well as open.

(iii) The projection $p: \tilde{X} \to X$ is an open mapping; that is, $p(\tilde{U})$ is open in X for every open $\tilde{U}$ in $\tilde{X}$.

These facts help us to visualize directly the covering space $(\tilde{X}, p)$.

1.3 Theorem (Lifting of Path). *Let $(\tilde{X}, p)$ be a covering space of X, $\tilde{x}_0 \in \tilde{X}$, $p(\tilde{x}_0) = x_0$, and $a: I \to X$ a path in X with $a(0) = x_0$. Then there exists one and only one path $\tilde{a}$ with $\tilde{x}_0 = \tilde{a}(0)$ such that $p \circ \tilde{a} = a$. The path $\tilde{a}$ is called a lifting of a, or more precisely, the lifting of a at $\tilde{x}_0$.*

1.4 Lemma. *Let $(\tilde{X}, p)$ be a covering space of X, and Y an arcwise connected and locally arcwise connected space. If the mappings f_0, f_1: $Y \to \tilde{X}$ are such that $p \circ f_0 = p \circ f_1$, then the set $Z = \{y \in Y: f_0(y) = f_1(y)\}$ is either empty or the same as Y.*

This lemma implies the uniqueness in the conclusions of both the preceding and the succeeding theorems.

1.5 Theorem (Lifting of Homotopy Between Paths).*Let $(\tilde{X}, p)$ be a covering space of $\tilde{X}$, and suppose that $\tilde{a}_0$ and $\tilde{a}_1$: $I \to \tilde{X}$ are paths in $\tilde{X}$ with the same starting point $\tilde{x}_0$. If $p \circ \tilde{a}_0 \fallingdotseq p \circ \tilde{a}_1$*

(homotopic with fixed endpoints), then $\tilde{a} \simeq \tilde{a}_1$, and, in particular, $\tilde{a}_0$ and a_1 have the same terminal point.

1.6 Corollary. *If $(\tilde{X}, p)$ is a covering space of X, then the cardinal number of the point sets $p^{-1}(x)$ is the same for all $x \in X$. This number is called the number of leaves of the covering space.*

1.7 Corollary. *Let $(\tilde{X}, p)$ be a covering space of X, $\tilde{x}_0 \in \tilde{X}$, and $\tilde{x}_0 = p(\tilde{x}_0)$. Then the homomorphism p_π: $\pi_1(\tilde{X}, \tilde{x}_0) \to \pi_1(X, x_0)$ induced by the projection p is injective.*

1.8 Corollary. *Let $(\tilde{X}, p)$ be a covering space of X. Then the subgroups $p_\pi(\pi_1(\tilde{X}, \tilde{x}))$ for all $\tilde{x} \in p^{-1}(x_0)$ are exactly a conjugacy class of subgroups of $\pi_1(X, x_0)$.*

1.9 Corollary. *Let $(\tilde{X}, p)$ be a covering space of X, $\tilde{x}_0 \in \tilde{X}$, and $x_0 = p(\tilde{x}_0)$. Let $H = p_\pi(\pi_1(\tilde{X}, \tilde{x}_0))$, a supgroup of $\pi_1(X, x_0)$, and again let $\{H\alpha\}$, $\alpha \in \pi_1(X, x_0)$ be the set of the right cosets of H in $\pi_1(X, x_0)$. Then there exists a one-one correspondence $\varphi:\{H\alpha\} \to p^{-1}(x_0)$. In particular, the cardinal number of $\{H\alpha\}$ is the same as the number of leaves of $(\tilde{X}, p)$.*

2. Two basic theorems on liftings of self-mappings of X

Theorem 1.3 in the last section is about the lifting of a path in X to a path in $(\tilde{X}, p)$, and Theorem 1.5 about the liftings of two homotopic paths with fixed endpoints to two homotopic paths with the same starting point. Now we shall turn from paths in X first to mappings f: $(Y, y_0) \to (X, x_0)$, that is, mappings f: $Y \to X$ with $f(y_0) = x_0$, where $y_0 \in Y$, $x_0 \in X$, and thus obtaining Lemmas 2.1 and 2.2; and then to self-mappings $f:(X, x_0) \to (X, f(x_0))$ thus obtaining finally Theorems 2.4 and 2.5 the two basic theorems being mentioned in the section title.

2.1 Lemma (Lifting of Mapping: $Y \to X$ to Mapping: $Y \to \tilde{X}$). *Let $(\tilde{X}, p)$ be a covering space of X, Y an arcwise connected and locally arcwise connected space, $y_0 \in Y$, $\tilde{x}_0 \in \tilde{X}$ and $x_0 = p(\tilde{x}_0)$. Given a mapping $f:(Y, y_0) \to (X, x_0)$, there exists a lifting*

$$\tilde{f}: (Y, y_0) \to (\tilde{X}, \tilde{x}_0),$$

that is, a mapping $\tilde{f}$ with $p \circ \tilde{f} = f$, if and only if

$$f_\pi(\pi_1(Y, y_0)) \subseteq p_\pi(\pi_1(\tilde{X}, \tilde{x}_0)).$$

Moreover, f is unique when it exists.

2.2 Lemma (Lifting of Homotopy Between Mappings: $Y \to X$ to that Between Mappings: $Y \to \tilde{X}$). *Let $(\tilde{X}, p)$ be a covering space of X, and Y an arcwise connected and locally arcwise connected space. Given a homotopy F: $Y \times I \to X$ with $F(\cdot, 0) = f_0$ and a lifting $\tilde{f}_0$: $Y \to \tilde{X}$ of f_0, there exists one and only one lifting $\tilde{F}$: $Y \times I \to \tilde{X}$ of F such that $\tilde{F}(\cdot, 0) = \tilde{f}_0$.*

Theorems 2.1 and 2.2 are respectively direct generalizations of Theorems 1.3 and 1.5. They will be used to derive Theorems 2.4 and 2.5 below, the theorems we need. Now we are to introduce in Definition 2.3 a different concept of liftings, the liftings of self-mappings of X to self-mappings of $\tilde{X}$.

2.3 Definition. Let $(\tilde{X}, p)$ be a covering space of X, f: $X \to X$ a self-mapping of X, and F: $X \times I \to X$ a homotopy of self-mappings of X. A self-mapping $\tilde{f}$: $\tilde{X} \to \tilde{X}$ is called a *lifting of the self-mapping* f of X, if

$$p \circ \tilde{f} = f \circ p.$$

A homotopy $\tilde{F}$: $\tilde{X} \times I \to \tilde{X}$ between self-mappings of $\tilde{X}$ is called a *lifting of the homotopy F between self-mappings* of X, if

$$p \circ \tilde{F}(\cdot, t) = F(\cdot, t) \circ p, \ \forall t \in I,$$

that is,

$$p \circ \tilde{F}(\tilde{x}, t) = F(p(\tilde{x}), t), \ \forall \tilde{x} \in \tilde{X}, \ \forall t \in I.$$

2.4 Theorem (Lifting of self-Mapping). *Let $(\tilde{X}, p)$ be a covering space of X, and f a self- mapping of X such that $f(x_0) = x_0'$. Let $\tilde{x}_0 \in p^{-1}(x_0)$, and $\tilde{x}_0' \in p^{-1}(x_0')$. Then there exists a lifting $\tilde{f}$: $(\tilde{X}, \tilde{x}_0) \to (\tilde{X}, \tilde{x}_0')$ when and only when*

$$f_\pi \circ p_\pi(\pi_1(\tilde{X}, \tilde{x}_0)) \subseteq p_\pi(\pi_1(\tilde{X}, \tilde{x}_0')).$$

Moreover, $\tilde{f}$ is unique when it exists.

2.5 Theorem (Lifting of Homotopy Between self-Mappings). *Let $(\tilde{X}, p)$ be a covering space of X, F: $X \times I \to X$ a homotopy between self-mappings of X, and $\tilde{f}_0$: $\tilde{X} \to \tilde{X}$ a lifting of $f_0 = F(\cdot, 0)$: $X \to X$. Then there exists a unique lifting $\tilde{F}$: $\tilde{X} \times I \to \tilde{X}$ such that $\tilde{f}_0 = \tilde{F}(\cdot, 0)$.*

3. Homomorphisms and automorphisms of covering spaces

The present section aims to obtain some information about the various possible covering spaces of a given space X.

3.1 Definition. Let $(\tilde{X}_1, p_1)$ and $(\tilde{X}_2, p_2)$be two covering spaces of X. If a mapping φ: $\tilde{X}_1 \rightarrow \tilde{X}_2$ is such that $p_2 \circ \varphi = p_1$,

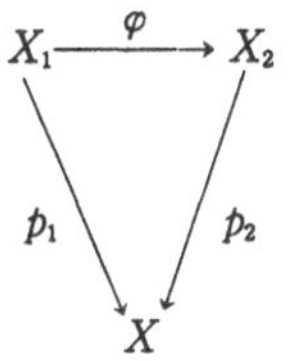

or

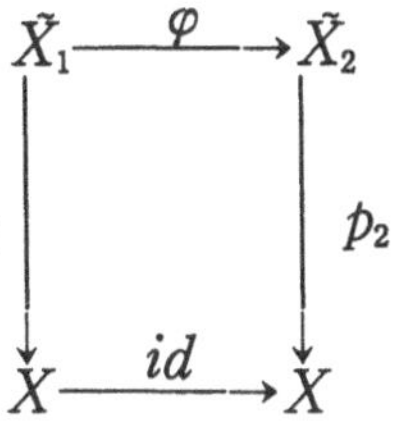

then φ is called a *homomorphisms* from $(\tilde{X}_1, p_1)$into $(\tilde{X}_2, p_2)$.

The product of two homomorphisms is obviously a homomorphism. The identity mapping id: $\tilde{X} \rightarrow \tilde{X}$ of a covering $(\tilde{X}, p)$ of X is also a homomorphism.

3.2 Definition. A homomorphism φ from $(\tilde{X}_1, p_1)$ into $(\tilde{X}_2, p_2)$ is called an *isomorphism,* if there exists a homomorphism ψ from (X_2, p_2) into $(\tilde{X}_1, p_1)$ such that $\psi \circ \varphi$ and $\varphi \circ \psi$ are both the identity mappings. When there is an isomorphism, the two covering spaces are *isomorphic.* A self-isomorphism of a covering space $(\tilde{X}, p)$ is called also a *covering motion,* which may or may not be the identity mapping, but must be a homeomorphism.

3.3 Definition. With the operation of composition of mappings, the totality of covering motions of a covering space $(\tilde{X}, p)$ of X is a group, called the *group of covering motions,* and is denoted by $\mathscr{D}(\tilde{X}, p)$.

Remark 1. In the theory of fibre bundles, the concept of fibre mapping is defined. For the special case of covering spaces $(\tilde{X}_i, p_i)$ of $X_i, i=1,2$, a mapping φ: $\tilde{X}_1 \rightarrow \tilde{X}_2$ is called a fibre mapping, when φ maps the fibre containing $\tilde{x}_1$ into the fibre containing $\varphi(\tilde{x}_1)$,for every $\tilde{x}_1 \in \tilde{X}_1$. It gives rise to a mapping f from X_1 to X_2 such that $p_2 \circ \varphi = f \circ p_1$. In the sense of Definition 2.3, φ may be called a lifting of the mapping f.

The homomorphism φ in Definition 3.1 is a fibre mapping for the case $X_1 = X_2 = X$ and $f = id$. The isomorphism φ in Definition 3.2 is a fibre mapping and at the same time a homeomorphism for this case. The covering motion is a fibre mapping and at the same time a homeomorphism for the case $X_1 = X_2 = X$, $(\tilde{X}_1, p_1) = (\tilde{X}_2, p_2)$ and $f = id$.

From Lemmas 1.4, 2.1 and Corollary 1.8, we can derive the following fundamental properties of homomorphisms, isomorphisms and covering motions of covering spaces.

3.4 Lemma. *Let φ_0 and φ_1 be homomorphisms from $(\tilde{X}_1, p_1)$ to $(\tilde{X}_2, p_2)$. If there exists any point $x \in \tilde{X}_1$ such that $\varphi_0(\tilde{x}) = \varphi_1(\tilde{x})$, then $\varphi_0 = \varphi_1$.*

3.5 Corollary. *If $\varphi \in \mathscr{D}(\tilde{X}, p)$ and $\varphi \neq id$, then φ has no fixed point.*

3.6 Lemma. *Let $(\tilde{X}_i, p_i)$ be covering spaces of X, $i=1,2$, and $\tilde{x}_i \in \tilde{X}_i$ such that $p_1(\tilde{x}_1) = p_2(\tilde{x}_2)$. Then there exists a homomorphism φ from $(\tilde{X}_1, p_1)$ to $(\tilde{X}_2, p_2)$ such that $\varphi(\tilde{x}_1) = \tilde{x}_2$, if and only if*

$$(p_1)_\pi(\pi_1(\tilde{X}_1, \tilde{x}_1)) \subseteq (p_2)_\pi(\pi_1(\tilde{X}_2, \tilde{x}_2)).$$

3.7 Corollary. *Under the hypotheses of Lemma 3.6, there exists an isomorphism φ from $(\tilde{X}_1, p_1)$ to $(\tilde{X}_2, p_2)$ such that $\varphi(\tilde{x}_1) = \tilde{x}_2$ if and only if*

$$(p_1)_\pi(\pi_1(\tilde{X}_1, \tilde{x}_1)) = (p_2)_\pi(\pi_1(\tilde{X}_2, \tilde{x}_2)).$$

3.8 Corollary. *Let $(\tilde{X}, p)$ be a covering space of X, and $\tilde{x}_1, \tilde{x}_2 \in p^{-1}(x_0)$, $x_0 \in X$. Then there exists a covering motion $\varphi \in \mathscr{D}(\tilde{X}, p)$ such that $\varphi(\tilde{x}_1) = \tilde{x}_2$, if and only if*

$$p_\pi(\pi_1(\tilde{X}, \tilde{x}_1)) = p_\pi(\pi_1(\tilde{X}, \tilde{x}_2)).$$

3.9 Theorem. *Two covering spaces $(\tilde{X}_i, p_i)$ of X are isomorphic, $i=1,2$, if and only if, for any two points $\tilde{x}_i \in \tilde{X}_i$ such that $p_i(\tilde{x}_i) = x_0$, the subgroups $(p_i)_\pi(\pi_1(\tilde{X}_i, \tilde{x}_i))$ belong to the same conjugacy class in $\pi_1 X, x_0)$.*

3.10 Lemma. *Let $(\tilde{X}_1, p_1)$ and $(\tilde{X}_2, p_2)$ be two covering spaces of X, and φ a homomorphism from the first covering space to the second. Then $(\tilde{X}_1, \varphi)$ is a covering space of $\tilde{X}_2$.*

Let $(\tilde{X}, p)$ be a simply connected covering space of X. If $(\tilde{X}', p')$ is another covering space of X, then from Lemma 3.6 there exists a homomorphism φ from $(\tilde{X}, p)$ to $(\tilde{X}', p')$, and from Lemma 3.10 $(\tilde{X}, \varphi)$ is a covering space of $\tilde{X}$. Thus $\tilde{X}$ is a covering space of any covering space $\tilde{X}'$ of X. This is why such $(\tilde{X}, p)$ is called an *universal covering space* of X. Furthermore, from Theorem 3.9, universal covering space is unique up to isomorphism.

When a covering space $(\tilde{X}, p)$ of X is such that $p_\pi(\pi_1(\tilde{X}, \tilde{x}_0))$ is a normal subgroup of $\pi_1(X, x_0)$, then $(\tilde{X}, p)$ is called a *regular covering space* of X. This definition is independent of the choice of the point $\tilde{x}_0 \in p^{-1}(x_0)$. The universal covering space of X is a regular covering space, because $\pi_1(\tilde{X}, \tilde{x}_0) = \{e\}$. From Corollary 3.8, we have the following:

3.11 Theorem. *Let $(\tilde{X}, p)$ be a regular covering space of X, and $\tilde{x}_1, \tilde{x}_2 \in p^{-1}(x_0)$, $x_0 \in X$. Then there exists a unique covering motion $\varphi \in \mathscr{D}(\tilde{X}, p)$ such that $\varphi(\tilde{x}_1) = \tilde{x}_2$.*

3.12 Theorem. *Let $(\tilde{X}, p)$ be a regular covering space of X, and $\tilde{x}_0 \in \tilde{X}$, $x_0 = p(\tilde{x}_0)$. Then, the group $\mathscr{D}(\tilde{X}, p)$ of covering motions is isomorphic to the quotient group $\pi_1(X, x_0)/p_\pi(\pi_1(\tilde{X}, \tilde{x}_0))$. In particular, if $(\tilde{X}, p)$ is the universal covering space, then $\mathscr{D}(\tilde{X}, p) \approx \pi_1(X, x_0)$.*

Remark 2. A generalization of Theorem 3.12. If $(\tilde{X}, p)$ is a general (not necessarily regular) covering space of X, then
$\mathscr{D}(\tilde{X}, p) \approx N[p_\pi(\pi_1(\tilde{X}, \tilde{x}_0))]/p_\pi(\pi_1(\tilde{X}, \tilde{x}_0))$, where $N[p_\pi(\pi_1(\tilde{X}, \tilde{x}_0))]$ is the normalizer of $p_\pi(\pi_1(\tilde{X}, \tilde{x}_0))$ in $\pi_1(X, x_0)$.

4. Geometrical construction of covering spaces

A topological space X is *locally simply connected,* if for each $x \in X$, there exists an open set U containing x such that any loop at x in U is homotopic in X (but not necessarily in U!) to the constant path or point path $e_x: I \to x$.

4.1 Theorem (Existence Theorem of Covering Space). *Let the topological space X be arcwise connected, locally arcwisee connected and locally simply connected, and a point $x_0 \in X$ be taken as the base point in X. Let H be a subgroup of the fundamental group $\pi_1(X, x_0)$ of X. Then there exists a covering space $(\tilde{X}, p)$ of X such that $p_\pi(\pi_1(\tilde{X}, \tilde{x}_0)) = H$, where $\tilde{x}_0 \in \tilde{X}$ and $p(\tilde{x}_0) = x_0$. In particular, if $H = \{e\}$, e being the identity element of $\pi_1(X, x_0)$, then $\tilde{X}$ is simply connected, and so each such X has a universal covering space.*

As a preliminary step for the next section,we shall sketch the idea of the proof, and at the same time introduce the notations needed.

A) Some properties of the abstractly defined covering space $\tilde{X}$

Let us ask first how the points of $\tilde{X}$ are related to X when $\tilde{X}$ is given abstractly as in § 1. If c and c' are paths in X starting from x_0 with $c \doteqdot c'$ (homotopic with fixed endpoints), then from Theorems 1.3 and 1.5 there exist a unique path $\tilde{c}$ and a unique path $\tilde{c}'$ in $\tilde{X}$ at $\tilde{x}_0$ such that $\tilde{c} \doteqdot \tilde{c}'$; and in particular $\tilde{c}(1) = \tilde{c}'(1)$. Thus the point $\tilde{c}(1)$ of $\tilde{X}$ corresponds to the path class $\langle c \rangle$ (by a path class we mean henceforth in this Appendix a path class with fixed endpoints) in X at x_0. Moreover, when we consider all the path classes in X at x_0 instead of only the loop classes at x_0, we see just as in Corollary 1.9 that the point $\tilde{c}$ (1) of $\tilde{X}$ corresponds to all the path classes

$$H\langle c\rangle = \{\langle h\rangle\langle c\rangle : \langle h\rangle \in H\}$$

(see Notation A 4.1 and Definition A 4.4), and find that the points of $\tilde{X}$ are in one-to-one correspondence with the path classes

$$\{H\gamma : \gamma \text{ a path class in } X \text{ at } x_0\}. \tag{1}$$

B) Geometrical construction

1) The point set $\tilde{X}$ and the projection p. The consideration in A) leads us to take the set (1) of path classes in X at x_0 as the point set of the covering space $\tilde{X}$ to be constructed. Define p: the point set $\tilde{X} \to X$ by $p(H\langle c\rangle) = c(1)$, and take $H\langle e_0\rangle$ as the base point $\tilde{x}_0$ of the covering space $\tilde{X}$ to be constructed, where $e_0: I \to x_0$ is the point path at x_0. Then the arcwise connectedness implies $p(\tilde{X}) = X$.

2) The topology in $\tilde{X}$. For convenience, an arcwise connected open set U of X, in which every loop is homotopic with fixed endpoints to a point path in X, will be called a *specified open set.* As a consequence of locally arcwise connectedness and locally simply connectedness, for any point $x \in X$ and any neighborhood V of x in X, there exists a specified open set U of X such that $x \in U \subseteq V$. Now let $H\langle c\rangle$ be a point of $\tilde{X}$, and U a specified open set of X such that $p(H\langle c\rangle) = c(1) \in U$. Form from the pair $H\langle c\rangle$ and U the following set in $\tilde{X}$ (Figure 1):

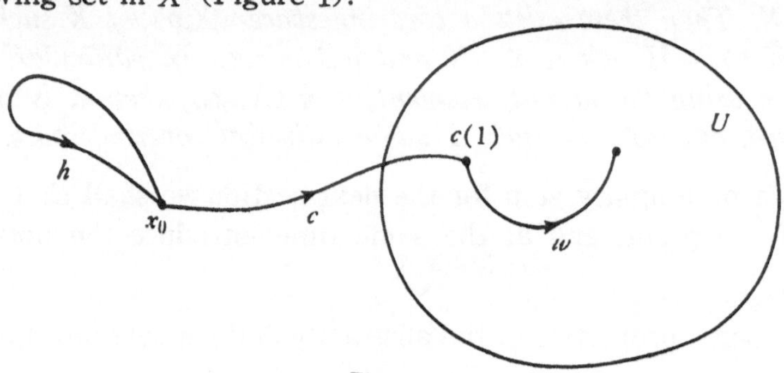

Figure 1

$$(H\langle c\rangle, U) = \{H\langle cw\rangle : w \text{ a path in } U \text{ starting from } c(1)\}$$

and call it a *specified open set of* $\tilde{X}$. It can be shown that the set of the specified open sets of the point set $\tilde{X}$ fulfills the requirement of a basis of a topological space, and thus defines with the point set $\tilde{X}$ a topological space $\tilde{X}$.

3) $\tilde{X}$ as an abstractly defined covering space of X. In order to conclude that the topological space $\tilde{X}$ obtained in 2) is indeed an abstractly defined covering space of X, we must prove that $p: \tilde{X} \to X$ is continuous, that the specified open sets of X are the admissible neighborhoods (Definition 1.2), and finally that $\tilde{X}$ is arcwise connected and locally arcwise connected. We omit all these proofs.

4) The proof of Theorem 4.1. The proof that $p_\pi(\pi_1(\tilde{X}, \tilde{x}_0)) = H$ in Theorem 4.1 is based on the following

4.2 Lemma (Lifting of Path). *Let c be a given path in X starting at x_0 and a any path in X starting at the terminal point of c. If the path $\tilde{a}$ in $\tilde{X}$ is defined (see Figure 2) by*

$$\tilde{a}: I \to \tilde{X},\ s \mapsto \tilde{a}(s) = H\ \langle ca_0^s \rangle\ ,\ 0 \leqslant s \leqslant 1,$$

then $\tilde{a}$ is the unique path in $\tilde{X}$ with $H\langle c \rangle$ and $H\langle ca \rangle$ as the starting and terminal points such that $p \circ \tilde{a} = a$.

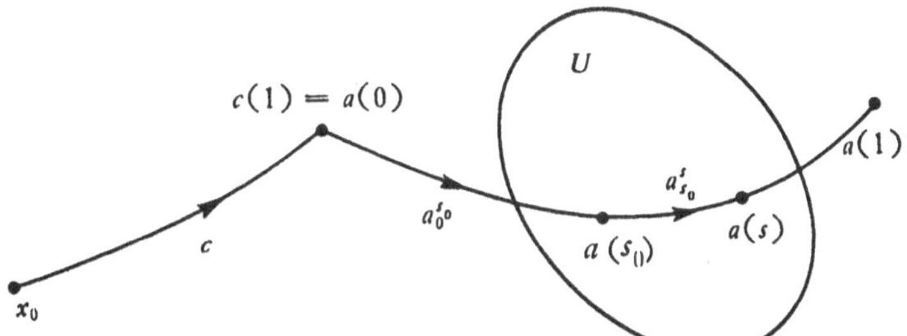

Figure 2

The reader can find a detailed proof of this theorem in [12]. We point out, however, that the notations in Figure 2 (which show the mapping $\tilde{a}$ is continuous at the point $s = s_0 \in I$) and in the geometrical formula of $\tilde{a}$ above are repeatedly used in the next section.

5. Geometrical formulas of liftings in the universal covering space

In this section we consider only the universal covering space $(\tilde{X}, p)$ of an arcwise connected, locally arcwise connected and locally simply connected space X, as constructed in § 4, and also only the liftings of self-mappings of X as given in Definition 2.3. Choose a point $x_0 \in X$ as the base point of X. Then take the point $\tilde{x}_0 = \langle e_0 \rangle \in \tilde{X}$ as the base point of $\tilde{X}$, where $e_0: I \to x_0$ is the point path at x_0. Because of $H = \{e\}$ for the universal covering space $\tilde{X}$, we have very simple geometrical formulas for the liftings in Theorems 2.4 and 2.5 as follows.

5.1 Theorem (Lifting of self-Mapping). *Let $(\tilde{X},p)$ be the universal covering space of X, and f a self-mapping of X, Take $\tilde{x}_0 = \langle e_0\rangle$, e_0 being the point path at x_0. Let w_0' be any path in X from the base point x_0 to $x_0' = f(x_0)$, and $\langle w_0'\rangle = \tilde{x}_0' \in \tilde{X}$. Then there exists a unique lifting $\tilde{f}: \tilde{X} \to \tilde{X}$ of f such that $\tilde{f}(\tilde{x}_0) = \tilde{x}_0'$. The geometrical formula of $\tilde{f}$ is*

$$\tilde{f}: \tilde{x} = \langle c_x\rangle \mapsto \tilde{x}' = \tilde{f}(\tilde{x}) = \langle w_0'(f \circ c_x)\rangle\ ,$$

where c_x denotes any path in X from x_0 to x.

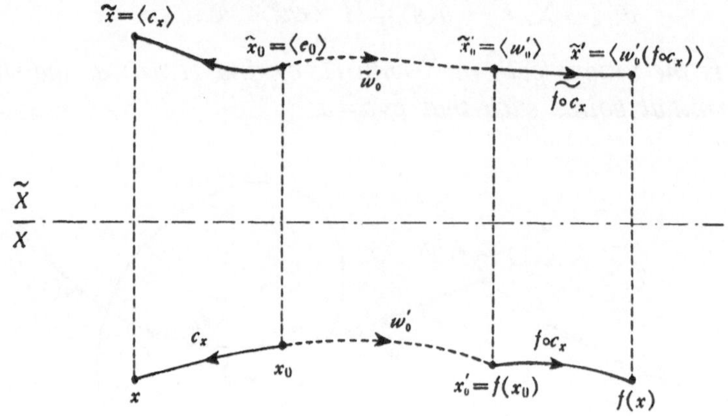

Figure 3

Proof. (see Figure 3) The existence and uniqueness of the lifting $\tilde{f}$ follows from Theorem 2.4. We only need to check the formula.

Consider in X the path c_x starting from x_0 and the path $f \circ c_x$ starting from $\tilde{x}_0' = f(x_0)$. From Lemma 4.2, the lifting of c_x starting from $\tilde{x}_0 = \langle e_0\rangle$ has as its terminal point the point $\langle e_0\, c_x\rangle = \langle c_x\rangle = \tilde{x}$, and the lifting of $f \circ c_x$ starting from $\tilde{x}_0' = \langle w_0'\rangle$ has as its terminal point the point $\langle w_0'(f \circ c_x)\rangle = \tilde{x}'$. Our formula of $\tilde{f}$ then follows from Definition 2.3. As far as we know, this formula appeared first in [36 II] § 3.

5.2 Corollary. (see Figure 4) *Let $\alpha = \langle a\rangle \in \pi_1(X, x_0)$. The lifting of the identity mapping of X, determined by $\tilde{x}_0 = \langle e_0\rangle \mapsto \tilde{x}_0' = \langle a\rangle$, is a covering motion of $\tilde{X}$, denoted also by α. The geometrical formula of the covering motion α is*

$$\alpha: \tilde{x} = \langle c_x\rangle \mapsto \tilde{x}' = \alpha(\tilde{x}) = \langle a c_x\rangle = \alpha\, \langle c_x\rangle\ .$$

It is worth while to note that for the geometrically constructed

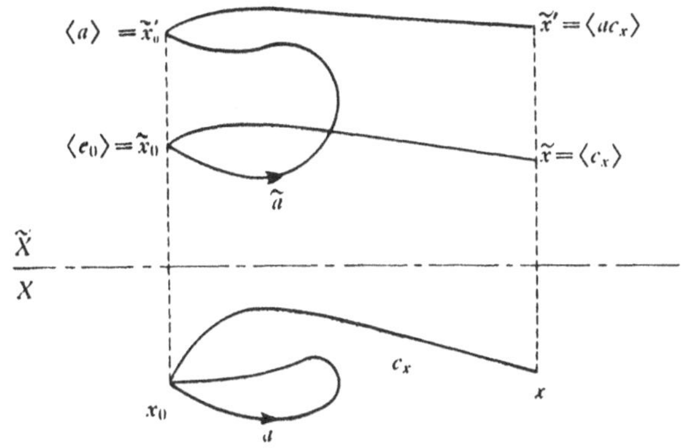

Figure 4

universal covering space, a covering motion α is given by left multiplication by α.

5.3 Corollary. (see Figure 5) *Let f be a self-mapping of X, and w_1, w_1' paths in X from x_0 to terminal points x_1 and $f(x_1)$ respectively. Then there exists a unique lifting $\tilde{f}$ of f such that $\tilde{f}$: $\tilde{x}_1 = \langle w_1 \rangle \mapsto \tilde{x}_1' = \langle w_1' \rangle$. The geometrical formula of $\tilde{f}$ is*

$$\tilde{f}:\ \tilde{x} = \langle c_x \rangle \mapsto \tilde{x}' = \tilde{f}(\tilde{x}) = \langle w_1' (f \circ w_1^{-1} c_x) \rangle \ .$$

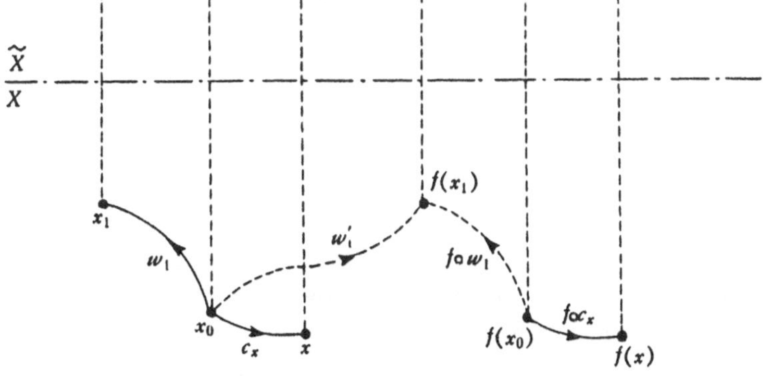

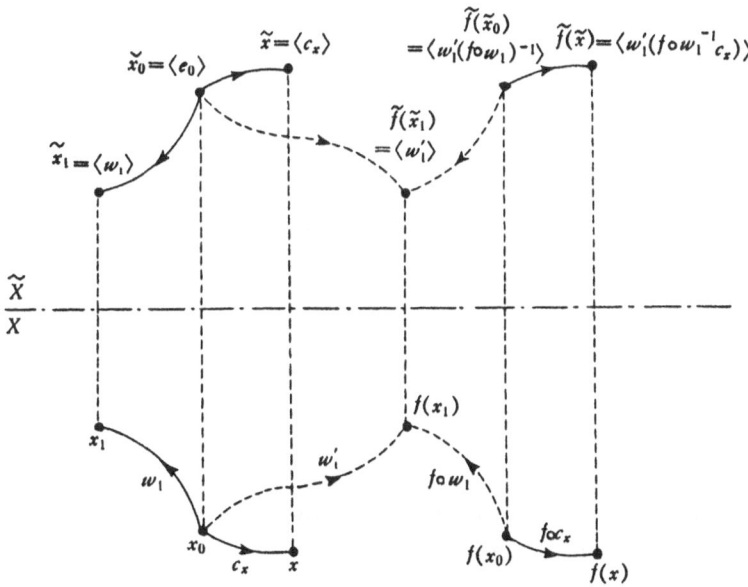

Figure 5

Proof. The geometrical formula requires checking. Let $\tilde{f}$: $\tilde{x}_0 = \langle e_0\rangle \mapsto \tilde{x}'_0 = \langle w'_0\rangle$, w'_0 being a path in X from x_0 to $f(x_0)$, a path to be determined. Then from Theorem 5.1, the geometrical formula of $\tilde{f}$ is

$$\tilde{f}: \tilde{x} = \langle c_x\rangle \mapsto \tilde{x}' = \langle w'_0(f\circ c_x)\rangle .$$

But the hypothesis $\tilde{f}$: $\langle w_1\rangle \mapsto \langle w'_1\rangle$ implies now $\langle w'_1\rangle = \langle w'_0\,(f\circ w_1)\rangle$, and therefore $\langle w'_0\rangle = \langle w'_1(f\circ w_1^{-1})\rangle$. Thus we obtain the geometrical formula desired.

5.4 Theorem (Lifting of Homotopy Between self-Mappings). (See Figure 6) *Let $(\tilde{X}, p)$ be the universal covering space of X, and f_t: $X\to X$, $t\in I$, a homotopy between the self-mappings $f = f_0$ and f_1 of X. Let x_0, e_0 and w'_0 with $x'_0 = f(x_0) = f_0(x_0)$, $\langle e_0\rangle = \tilde{x}_0$, $\langle w'_0\rangle = \tilde{x}'_0$ be the same as in Theorem* 5.1. *Then there exists a unique lifting $\tilde{f}_t$: $\tilde{X}\to\tilde{X}$ of f_t such that $\tilde{f}_0(\tilde{x}_0) = \tilde{x}'_0$. The geometrical formula of $\tilde{f}_t$ is*

$$\tilde{f}_t: \tilde{x} = \langle c_x\rangle \mapsto \tilde{x}'_t = \tilde{f}_t\,(\tilde{x}) = \langle w'_0 b_0^t\,(f_t\circ c_x)\rangle ,$$

where b_0^t is the subpath (Definition A 2.3) *of $b = f_t\,(x_0)$ in X.*

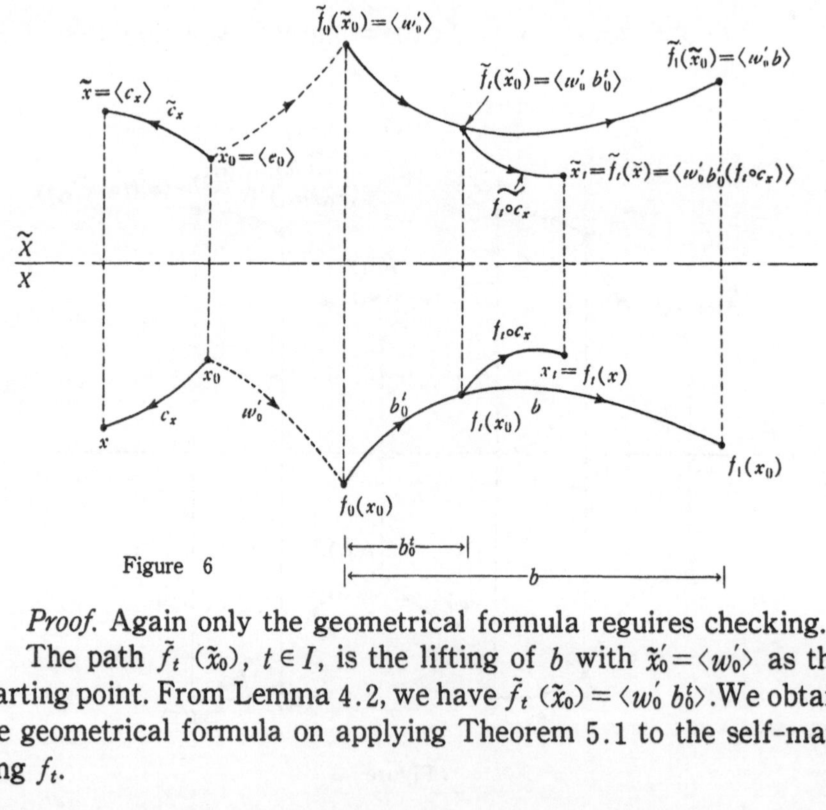

Figure 6

Proof. Again only the geometrical formula reguires checking.

The path $\tilde{f}_t\,(\tilde{x}_0)$, $t\in I$, is the lifting of b with $\tilde{x}'_0 = \langle w'_0\rangle$ as the starting point. From Lemma 4.2, we have $\tilde{f}_t\,(\tilde{x}_0) = \langle w'_0\; b_0^t\rangle$. We obtain the geometrical formula on applying Theorem 5.1 to the self-mapping f_t.

Appendix C

APPROXIMATION THEOREMS

Let K be as usual a connected finite simplicial complex, and f a self -mapping of the polyhedron $|K|$. When the least number of fixed points of the mapping class $\langle f \rangle$ of f is under Investigation, one needs to modify f repeatedly by means of the so-called "short homotopy" to secure a new homotopic self-mapping g such that g differs from f only on a subset U of $|K|$ and that the fixed point set of g in U possesses certain specified property. In § 1, the construction and some basic properties of the short homotopy will be given first (Theorem 1.1) and then we shall show in general that there exists a short homotopy between f and g on U when $f(x)$ and $g(x)$ are "very near" $\forall\ x \in U$ (Theorem 1.5). Making use of § 1, we can proceed in § 2 to show that for any given f and certain-choice of a subset U of $|K|$, there always exists a new homotopic self-mapping g of $|K|$ such that g differs from f only on U but is still very near to f there, and that all the fixed points of g in $|K|$ are isolated and g is linear in a neighborhood of every fixed point (Theorems 2.5 and 2.9).

This appendix is prepared as a prerequisite for the operations of "moving" and "uniting" the fixed points in Chapter IV. In this appendix we always use closed simplexes and denote them by $\underline{t}, \underline{s}, \underline{\sigma}$ etc.

1. Short homotopy between self-mappings

Every point $x \in |K|$ lies in a unique closed simplex of the lowest dimension in K. The closed simplex is called the closed *carrier* of x and is denoted by $\mathrm{Car}_K x$, or by $\mathrm{Car}\,x$ for short when no ambiguity arises. The open carrier of x is similarly defined, and is denoted by $\mathrm{car}_K x$ or $\mathrm{car}\,x$.

Let x and y be any two points of $|K|$. If their closed carriers have nonempty intersection, i. e.

$$\mathrm{Car}\,x \cap \mathrm{Car}\,y \neq \emptyset,$$

then the intersection is a closed simplex of K, the common face of Car x and Car y. Thus, for any point z on this face, the broken-line $[x, z, y]$ is completely determined and is on $|K|$. For such a given pair x and y, how is a suitable point z chosen so that the broken-line $[x, z, y]$ may vary in a certain continuous manner with the pair? The choice is described in the following theorem.

1.1 Theorem. *Let K be a connected finite simplicial complex. Define*

$$M = \{(x, y) \in |K| \times |K| \colon \mathrm{Car}\,x \cap \mathrm{Car}\,y \neq \emptyset\},$$

which is an open subset on $|K| \times |K|$. Then there exists a mapping

$$\alpha\colon M \times I \to |K|,\ (x, y, t) \mapsto \alpha(x, y, t),$$

(called α-deformation) with the following properties:

(i) $\alpha(x, y, 0) = x,$ *when* $(x, y) \in M$;
(ii) $\alpha(x, y, 1) = y,$ *when* $(x, y) \in M$;
(iii) $\alpha(x, x, t) = x,$ *when* $x \in |K|, 0 \leqslant t \leqslant 1$;
(iv) $\mathrm{Car}\,x \cap \mathrm{Car}\,\alpha(x, y, t) \neq \emptyset,$ *when* $(x, y) \in M, 0 \leqslant t \leqslant 1$;
(v) $\alpha(x, y, t) \neq x,$ *when* $(x, y) \in M, x \neq y, 0 < t \leqslant 1$;
(vi) *For any given number $\varepsilon > 0$, there exists a number $\delta > 0$ such that $d(x, y) < \delta$ implies both $(x,y) \in M$ and*

$$d(x, \alpha(x, y, t)) < \varepsilon,\ \textit{when}\ 0 \leqslant t \leqslant 1.$$

Proof. Definition of the α-deformation is briefly given as follows. Arrange the vertices of K in a sequence

$$a_0, a_1, \cdots, a_n,$$

which may be supposed to span an n-simplex in R^n, and K may be regarded as a subcomplex of this n-simplex. Then any point $x \in |K|$ has in K its barycentric coordinates

$$\lambda_0, \lambda_1, \cdots, \lambda_n$$

($\lambda_i \geqslant 0$, $i = 0,1,\cdots,n$, and $\sum_{i=0}^{n} \lambda_i = 1$),and the vertices of Car$x$ are just those a_i corresponding to positive λ_i. Let $y \in |K|$ have barycentric corrdinates λ'_i, $i = 0,1,\cdots,n$. Set

$$\beta(x, y) = \sum_{i=0}^{n} \sqrt{\lambda_i \lambda_i'}.$$

$\beta(x,y) > 0$ if and only if $(x,y) \in M$.

Now, for $(x,y) \in M$, set

$$\lambda_i'' = \sqrt{\lambda_i \lambda_i'} / \beta(x, y).$$

$\lambda_0'', \lambda_1'', \cdots, \lambda_n''$ are the barycentric coordinates of a definite point z on $\mathrm{Car}x \cap \mathrm{Car}y$, and the function

$$z = z\ (x, y) = (\lambda_0'', \lambda_1'', \cdots, \lambda_n''),\ \forall (x, y) \in M,$$

is thus a single-valued continuous function. Via z there is uniquely defined on $|K|$ the broken-line $[x, z, y]$, which degenerates into the single point x when and only when $x = y$. Finally define

$$\alpha(x, y, t),\ 0 \leqslant t \leqslant 1,$$

as the point on the broken-line, whose distance from x measured along the broken-line is t times the total length of the broken-line.

The α-deformation is a generalization of [3], Example III 3.3 or [8], Example II 10.3 which appeared in [36 I] p. 665, [36 III] p. 552. Our Theorem 1.1 without the properties (iv) and (vi) is [35], Lemma 1.1 and [2] VIII C, Lemma 1. (iv) is rather obvious, and we only need to indicate the proof of (vi).

For any given $\varepsilon > 0$, consider in M the subset

$$W = \{(x, y) \in M: d(x, z) + d(z, y) < \varepsilon\}.$$

Because of the continuity of $z = z\ (x, y)$, W is open in M, and hence open also in $|K| \times |K|$. On the other hand, W contains the diagonal Δ of $|K| \times |K|$, a closed subset of $|K| \times |K|$. Thus

$$|K| \times |K| \supseteq M \supseteq W \supseteq \Delta.$$

Let δ be the distance between Δ and $|K| \times |K| - W$. Then

$$d(x, y) < \delta \Rightarrow d((x, y), (x, x)) < \delta \Rightarrow (x, y) \in W.$$

Now $(x, y) \in W \Rightarrow (x, y) \in M$ as well as $d(x, z) + d\ (z, y) < \varepsilon$; and hence

$$d(x, \alpha(x, y, t)) < \varepsilon \text{ when } 0 \leqslant t \leqslant 1. \qquad \square$$

1.2 Definition. Let K be a simplicial complex, U a topological space, and f, g: $U \to |K|$ mappings. If

$$\mathrm{Car}f(y) \cap \mathrm{Car}g(y) \neq \emptyset,\ \forall y \in U,$$

then we say *f and g possess the property of short homotopy on U*,or briefly the *property S on U.*

The defining property in Definition 1.2 means just $(f(y), g(y)) \in M$, $\forall\ y \in U$, where M is defined in Theorem 1.1. Because M is an open subset of $|K| \times |K|$, the subset of U on which f and g possess the property S, must be also open in U. This fact will be referred to frequently.

1.3 Definition. When in Definition 1.2 the mapping $g: U \to |K|$ is the identity mapping id, then we say *f possesses the property S on U.*

Regarding property S, we point out first the following

1.4 Theorem. *Let f, g:* $U \to |K|$ *be the same as in Definition* 1.2. *There exists a number* $\delta > 0$ *such that*

$$d(f(y), g(y)) < \delta, \ \forall y \in U,$$

$\Rightarrow$*f and g possess the property S on U.*

Rroof. The property (vi) in Theorem 1.1 implies our theorem. But the following proof is simpler.

The open stars of all the vertices of K form an open covering of $|K|$. The Lebesgue number of this covering is such a number δ. □

Combining the above two definitions with (vi) in Theorem 1.1, we easily have the following results convenient for our use.

1.5 Theorem. *Let f, g:* $U \to |K|$ *be the same as in Definition* 1.2. *If f and g possess the property S on U, then there exists on U a homotopy* f_t: $f \simeq g: U \to |K|$ *such that* $f(y) = g(y)$, $y \in U$, *implies* $f_t(y) = f(y)$, *i. e., (see footnote* 1) *in Lemma IV* 1.1)

$$f_t: f \simeq g: U \to |K| \text{ rel } \{y \in U: f(y) = g(y)\}.$$

1.6 Corollary. *Let f, id:* $U \to |K|$ *be the same as in Definition* 1.3. *If f possesses the property S on U, then there exists on U a homotopy*

$$f_t: f \simeq id: U \to |K| \text{ rel } \Phi(f).$$

2. Approximation theorems

The following approximation theorem is classical. For its proof, see for instance [3], Theorem III 3.6 or [8], Theorem IV 5.8.

2.1 Theorem (The Simplicial Approximation Theorem). *Let K and L be two connected finite simplicial complexes and $f: |K| \to |L|$ a mapping. There exists a simplicial mapping g from a barycentric subdivision* $\mathrm{Sd}^{(r)} K$ *of K of certain order $r \geqslant 0$ to L:*

$$g: \mathrm{Sd}^{(r)} K \to L,$$

such that $g(x) \in \mathrm{Car}_L f(x)$, $\forall x \in |K|$. Thus $g \simeq f: |K| \to |I|$, and

$$d(f(x), g(x)) \leqslant \text{mesh } L, \ \forall x \in |K|,$$

where mesh L *denotes the maximum of the diameters of the simplexes L.*

When fixed points of self-mappings of polyhedra are investigated by means of simplicial approximation, we need the following definitions which play important role in another powerful approximate theorem, Theorem 2.5.

2.2 Definition. Let L be a connected finite simplicial complex, and $\underline{t}$, a closed simplex of L. According as $\underline{t}$ is not or is a face of a simplex of higher dimension of L, $\underline{t}$ is called a *maximal* or a *non-maximal simplex* of L. If $\underline{s}$ is a simplex of a barycentric subdivision $\mathrm{Sd}^{(r)} L$, $r \geqslant 0$, the simplex of the lowest dimension of L, which contains $\underline{s}$ is called the *carrier* of $\underline{s}$ in L.

2.3 Definition. Let L be a connected finite simplicial complex, K a connected subcomplex of the barycentric subdivision $\mathrm{Sd}^{(r)} L$, $r \geqslant 0$, and

$$g: K \to L$$

a simplicial mapping. If $\underline{s} \subseteq g(\underline{s})$, $\underline{s} \in K$, then $\underline{s}$ is called a *fixed simplex* of g. According as the fixed simplex $\underline{s}$ of g is a maximal or a non-maximal simplex of $\mathrm{Sd}^{(r)} L$, it is called a *maximal* or a *non-maximal fixed simplex* of g.

Let us point out here that by a simplex we always mean a simplex of dimension $\geqslant 0$. Moreover, if the fixed simplex $\underline{s}$ of g in Definition 2.3 is a maximal simplex of K but not a maximal simplex of $\mathrm{Sd}^{(r)} L$, then it is by definition a non-maximal fixed simplex of g. Some simple facts about fixed simplexes are given in the following

2.4 Lemma. *Let the notations be the same as in Definition* 2.3.

(i) *If $\underline{s}^p$ is a p-dimensional fixed simplex of g, $p \geqslant 0$, then $g(\underline{s}^p)$ is a p-dimensional simplex of L and is also the carrier of $\underline{s}^p$ in L.*

(ii) *If a point $x \in |K|$ is a fixed point of g and the simplex $\underline{s}$ is the carrier of x in K, then $\underline{s}$ is a fixed simplex of g.*

(iii) *If $\underline{s}$ is a fixed simplex of g, then g has at least one fixed point in $\underline{s}$.*

(iv) *If a simplex $\underline{s}$ is a maximal fixed simplex of g and each of its proper faces is not a fixed simplex of g, then the fixed point of g in $\underline{s}$ is unique and is an isolated fixed point.*

(v) *If the fixed simplex $\underline{s}$ of g is not a maximal fixed simplex, then $g(\underline{s})$ is not a maximal simplex of L.*

Proof. The proofs of (i) and (ii) are left to the reader.

(iii) From the hypothesis and (i), $g|\underline{s}$, is a non-degenerate linear mapping ([1], p. 174), and hence $(g|\underline{s})^{-1}$: $g(s) \to \underline{s} \subseteq g(\underline{s})$ is a mapping and has at least one fixed point by the Brouwer fixed point theorem. Thus $g|\underline{s}$ has at least one fixed point.

(iv) From the hypothesis and (ii), g has no fixed point on the boundary of $\underline{s}$. From (iii), g has at least one fixed point $x \in \underline{s}$. Because of the non- degenerate linear property of g on $\underline{s}$, g can not have another fixed point $x' \neq x$ in $\underline{s}$, or else every point of the straight line xx', and in particular the point of intersection of xx' with the boundary of $\underline{s}$, would be a fixed point of g. Thus g has a unique fixed point in $\underline{s}$. Moreover, since $\underline{s}$ is a maximal simplex of $\mathrm{Sd}^{(r)}\, L$, the interior of $\underline{s}$ is an open subset of $|L|$. Hence the fixed point x of g is isolated.

(v) Let the p-dimensional fixed simplex $\underline{s}^p$ be a proper face of a $(p+1)$-dimensional simplex $\underline{s}^{p+1}$ of $\mathrm{Sd}^{(r)}\, L$, and $\underline{t}$ the carrier of $\underline{s}^{p+1}$ in L. Then $\underline{s}^p \subseteq \underline{s}^{p+1} \subseteq \underline{t}$. From (i), $g(\underline{s}^p)$ is the carrier of $\underline{s}^p$ in L, and hence $g(\underline{s}^p)$ is a face of $\underline{t}$. Moreover, dim $\underline{t} \geqslant p+1$. Thus $g(\underline{s}^p)$ is a proper face of $\underline{t}$. $\square$

Now we are ready to prove the Hopf approximation theorem ([1], p. 543) in the following more general form, the main theorem in the present section. It is more general in the sense that the mapping considered is no longer limited to a self-mapping of a polyhedron of a homogeneous n-dimensional complex, but is from a subpolyhedron of a general polyhedron to the polyhedron.

2.5 Theorem (The Hopf Approximation Theorem). *Let L be a connected finite simplicial complex, K a connected subcomplex of L and f: $|K| \to |L|$ a mapping. For any given number $\varepsilon > 0$, there exists a simplicial mapping*

$$g: \mathrm{Sd}^{(k)} K \to \mathrm{Sd}^{(l)} L$$

from a subdivision of K of certain order k to a subdivision of L of a certain order $l(\leqslant k)$ such that

(i) *all the fixed simplexes of* g *are maximal fixed simplexes, and hence* g *has only a finite number of fixed points, and is linear in a neighborhood of each fixed point;*

(ii) *there exists a homotopy* $g_t: f \simeq g: |K| \to |L|$ *such that* $d(g_t(x), f(x)) < \varepsilon$, $\forall (x, t) \in |K| \times I$.

For the proof of this important theorem, it is necessary to make the following technical preliminaries.

2.6 Definition. Let L be a connected finite simplicial complex and K a subcomplex of the subdivision $\mathrm{Sd}^{(r)} L$ of order r of L, $r \geqslant 1$. Every simplex $\underline{s}^q$ of dimension q of K can be uniquely represented in the form

$$\underline{s}^q = (t_0^*, \ t_1^*, \cdots, \ t_q^*),$$

where $\underline{t}_i$ is a simplex of $\mathrm{Sd}^{(r-1)} L$, $i = 0,1 \cdots, q$, $\underline{t}_j$ stands for a proper face of $\underline{t}_{j+1}$, $j = 0,1, \cdots, q-1$, and t_i^* denotes the barycenter of $\underline{t}_i$. This representation of $\underline{s}^q$ is called the *barycentric representation.*

2.7 Lemma. *Let* K, L *and* $\underline{s}^q \in K$ *be the same as in Definition* 2.6. *If a* p*-dimensional face* $\underline{s}^q$ *of* $\underline{s}^q$, $p \leqslant q$, *is a fixed simplex of a simplicial mapping* $g: K \to L$, *then it has the barycentric representation*

$$\underline{s}^p = (t_0^*, \ t_1^*, \ \cdots, \ t_p^*),$$

where $\underline{t}_i$ *is* i*-dimensional,* $i = 0, 1, \cdots, p$.

Proof. From the hypothesis and Lemma 2.4(i), $\underline{s}^p$ is a fixed simplex of $g \Rightarrow \dim g(\underline{s}^p) = p$ and $\underline{s}^p \subseteq g(\underline{s}^p) \subseteq |L^p|$, where L^p is the p-dimensional skeleton of L.

$t_i^* \in |L^p| \Rightarrow \dim \underline{t}_i \leqslant p$. Since $\dim \underline{t}_i$ increases with i monotonically and $\underline{s}^p$ is p-dimensional, the conclusion follows at once. □

The proof of Theorem 2.5 is based on the following construction, a generalization of the so-called the Hopf construction.

2.8 Lemma. *Let* L *be a connected finite simplicial complex and* K *a subcomplex of* $\mathrm{Sd}^{(r)} L$, $r \geqslant 1$. *If* $g: K \to L$ *is a simplicial mapping, whose fixed simplexes of dimensions* $< p$ $(p \geqslant 0)$ *are all maximal fixed simplexes, then there exists a simplicial mapping* g': $\mathrm{Sd}\, K \to L$, *whose fixed simplexes of dimensions* $\leqslant p$ *are all maximal fixed simplexes, and*

$$d(g'(x), g(x)) \leqslant 2 \text{ mesh } L, \ \forall x \in |K|.$$

Proof. 1) Let the non-maximal fixed simplexes of dimension p of g be $\underline{\sigma}_1, \underline{\sigma}_2, \cdots, \underline{\sigma}'_\mu$. Construct first a correspondence g' from the vertices of Sd K, i. e. the set $\{\overset{*}{B} : \underline{s} \in K\}$, to those of L as follows.

1 a) If $\underline{s}$ is a certain $\underline{\sigma}_j$, then from Lemma 2.4(v), $g(\underline{\sigma}_j)$ is not a maximal simplex of L, that is, $g(\underline{\sigma}_j)$ is a proper face of a simplex $\underline{\tau}_j$ of L. Take a vertex b_j of $\underline{\tau}_j$ which is not a vertex of $g(\underline{\sigma}_j)$, and define $g'(\overset{*}{s}) = b_j$.

1 b) If $\underline{s}$ has a certain $\underline{\sigma}_j$ as a proper face, then from Lemma 2.7, there is only one such $\underline{\sigma}_j$. Take any definite vertex of $g(\underline{\sigma}_j)$, and set it $= g'(\overset{*}{s})$.

1 c) If $\underline{s}$ is otherwise, take any definite vertex of g $(\underline{s})$ and set it $= g'(\overset{*}{s})$.

Example on the construction of g′ from g

Let L be the 2-simplex uvw, $K = um \in \text{Sd } L$, and g: $K \to L$ given by $g(u) = u$, $g(m) = v$, with u as the only non-maximal fixed simplex of dimension zero of g. By means of 1 a), 1 b) and 1 c) there are two different constructions g'_1 and g'_2 of g': Sd $K \to L$ as follows:

(i) Take $\underline{\tau}_j = uv$. Then $g'_1(u) = v$, $g'_1(n) = u$, $g'_1(m) = v$.
(ii) Take $\underline{\tau}_j = vw$. Then $g'_2(u) = w$, $g'_2(n) = u$, $g'_2(m) = v$.

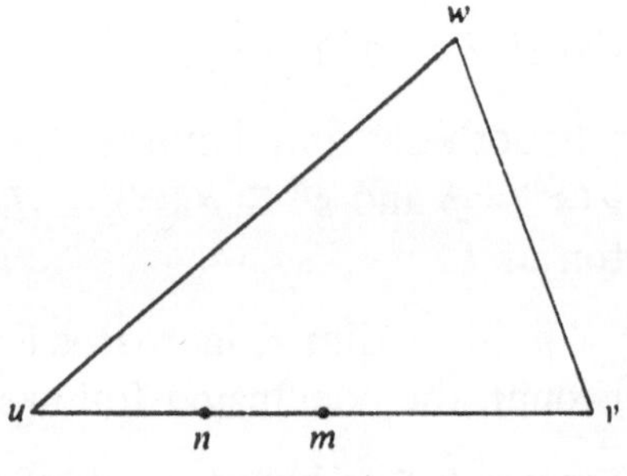

Although there is another possible choice of $\underline{\tau}_j$, namely $\underline{\tau}_j = uvw$, then $g'(u)$ is either v or w again and g' must be respectively g'_1 or g'_2 just obtained above.

Both g'_1 and g'_2 now no longer have non-maximal fixed 0-simplex, but g'_1 has two non-maximal fixed 1-simplexes un and nm, and g'_2 has one non-maximal fixed 1-simplex nm. Let us construct first from the g'_1: Sd$K \to L$ in (i) the desired new

$$g''_1 : \text{Sd}^{(2)}K \to L.$$

Let the barycenters of un and nm be a and b respectively. Then from la), we have $g''_1(a) = g''_1(b) = w$, and from 1 c) we have $g''_1(u) = v$, $g''_1(n) = u$, $g''_1(m) = v$, g''_1 has no longer any non-maximal fixed simplex.

We advise the reader to construct g''_2 from g'_2. The reader will find that there are again two possibilities: g_2'' can carry the barycenter a either to w or to u.

2) Prove that the correspondence g': $\mathrm{Sd}K \to L$ defined above carries all the vertices of every simplex of Sd K into the vertices of a certain simplex of L, and thus determines a simplicial mapping which will be denoted also by g'.

Let $\underline{s}' = (\overset{*}{s}_0, \overset{*}{s}_1, \cdots, \overset{*}{s}_q)$ be the barycentric representation of a simplex of Sd K where $\underline{s}_i$ are simplexes of K, $i = 0, 1, \cdots, q$. The simplexes of Sd K are of two different types.

$\underline{s}'$ is called a simplex of *the first type,* if among these $\underline{s}_i'$ there is no p-dimensional non-maximal fixed simplex of g. For such $\underline{s}_i'$ (cf, 1 b) and 1 c) above), there exists $g'(\overset{*}{s}_i) \in g(\underline{s}_i) \subseteq g(\underline{s}_q)$. Hence $g'(\underline{s}') \subseteq g(\underline{s}_q)$.

$\underline{s}'$ is called a simplex of Sd K of *the second type,* if it is not of the first type, that is, if among these $\underline{s}_i$ there is a certain $\underline{\sigma}_j$. Since $\underline{s}'$ is in barycentric representation, among these $\underline{s}_i$ there can be only one such p-dimensional simplex $\underline{s}_m = \underline{\sigma}_j$. When $i<$m, from 1 c), we have

$$g'(\overset{*}{s}_i) \in g(\underline{s}_i) \subseteq g(\underline{s}_m) \subseteq \underline{\tau}_j;$$

when $i = m$, from 1 a) we have $g'(\overset{*}{s}_m) = b_j \in \underline{\tau}_j$; and finally when $i>m$, from 1 b) we have $g'(\overset{*}{s}_i) \in g(\underline{\sigma}_j) \subseteq \underline{\tau}_j$. Hence $g'(\underline{s}') \subseteq \underline{\tau}_j$ for $\underline{s}'$ of the second type.

Thus the correspondence g' determines a simplicial mapping denoted also by g'.

3) Prove that all fixed simplexes of dimensions $\leqslant p$ $(p \geqslant 0)$ of g' are maximal fixed simplexes.

Let $\underline{s}' = (\overset{*}{s}_0, \overset{*}{s}_1, \cdots, \overset{*}{s}_q)$ be a q-dimensional fixed simplex of Sd K, $q \leqslant p$. For such $\underline{s}'$ the following two facts hold: (a) In the barycentric representation of $\underline{s}'$, dim $\underline{s}_i = i$, $i = 0, 1, \cdots, q$. (This follows when Lemma 2.7 is applied to Sd K.) (b) $g'(\underline{s}')$ is the carrier of $\underline{s}_q$ in L. (First, from definition of the barycentric representation, $\underline{s}_q$ is the carrier of $\underline{s}'$ in K as well as in $\mathrm{Sd}^{(r)}L$. Second from Lemma 2.4 (i), $g'(\underline{s}')$ is the carrier of $\underline{s}'$in L. Thus there follows the assertion (b).)

We say that $\underline{s}'$ can not be a simplex of the second type defined in 2). On the contrary if $\underline{s}'$ were a simplex of the second type, then from (a) and the hypothesis $q \leqslant p$ we would have $q = p$ and $\underline{s}_q =$ certain $\underline{\sigma}_j$; since $\underline{\sigma}_j$ is a fixed simplex of g, from Lemma 2.4 (i), $g(\underline{\sigma}_j)$ would be the carrier of $\underline{s}_q$ in L; and finally from (b) we would have $g'(\underline{s}') = g(\underline{\sigma}_j)$. But according to the construction of g', $g'(\underline{s}')$

contains a vertex b_j which is not a vertex of $g(\underline{\sigma}_j)$. This contradiction proves our assertion.

Thus $\underline{s}'$ must be a simplex of the first type. From 2), $g'(\underline{s}') \subseteq g(\underline{s}_q)$. This together with (b)$\Rightarrow \underline{s}_q \subseteq g(\underline{s}_q)$, i. e., $\underline{s}_q$ is a fixed simplex of g. From (a), the dimension of $\underline{s}_q$ is $q \leqslant p$. Since by hypothesis, all fixed simplexes of dimensions $q \leqslant p$ of g are maximal fixed simplexes of g, except the p-dimensional $\underline{\sigma}_1, \underline{\sigma}_2, \cdots, \underline{\sigma}_\mu$, then $\underline{s}_q$ is a maximal fixed simplex of g, and is a maximal simplex of $\mathrm{Sd}^{(r)} L$. From (a) again, we conclude that $\underline{s}'$ is a maximal simplex of $\mathrm{Sd}^{(r+1)} L$. Thus $\underline{s}'$ is a maximal fixed simplex of g'.

4) Prove that $d(g'(x),g(x)) \leqslant 2$ mesh L.

From the construction of g' from g in 2), we observe that if $\underline{s}'$ is a simplex of the first type of Sd K, then $g'(\underline{s}') \subseteq g(\underline{s}_q)$ and at the same time $g(\underline{s}') \subseteq g(\underline{s}_q)$. Thus $d(g'(x), g(x)) \leqslant$ mesh L, $\forall x \in \underline{s}'$ of the first type. If $\underline{s}'$ is a simplex of the second type of Sd K, then $g'(\underline{s}') \subseteq \underline{\tau}_j$ and $g(\underline{s}') \subseteq g(\underline{s}_q)$, that means that the simplexes $\underline{\tau}_j$ and $g(\underline{s}_q)$ have a common face $g(\underline{\sigma}_j)$. Thus $d(g'(x), g(x)) \leqslant 2$ mesh L, $\forall x \in \underline{s}'$ of the second type. □

Proof of Theorem 2.5. Let dim $L = n$. Take a subdivision $L_0 = \mathrm{Sd}^{(l)} L$ of L of an order l, which is to be determined later on. From simplicial approximation Theorem 2.1, given the mapping $f\colon |K| \to |L|$, there exist a subdivision $K_0 = \mathrm{Sd}^{(k_0)} K$, of order $K_0 > l$, and a simplicial mapping $g_0\colon K_0 \to L_0$, such that $g_0 \simeq f\colon |K| \to |L|$ and that $d(g_0(x), f(x)) \leqslant$ mesh L_0, $\forall x \in |K|$.

On applying Lemma 2.8 repeatedly we obtain after $g_0\colon K_0 \to L_0$ a sequence of simplicial mappings $g_p\colon K_p = \mathrm{Sd}^{(p)} K_0 \to L_0$, $p = 1,2,\cdots, n$, such that the fixed simplexes of g_p of dimensions $< p$ are all maximal fixed simplexes and that $d(g_p(x), g_{p-1}(x)) \leqslant 2$ mesh L_0. Let us note here that we regard g_0 as without fixed simplex of negative dimension and thus g_0 satisfies the hypothesis of Lemma 2.8.

Set $k = k_0 + n$ and $g = g_n\colon \mathrm{Sd}^{(k)} K \to \mathrm{Sd}^{(l)} L$. Then g satisfies (i) and $d(g(x), f(x)) \leqslant (2n+1)$mesh L_0, $\forall x \in |K|$.

According to Theorem 1.1(vi), for any given $\varepsilon > 0$, there exists a number $\delta > 0$ such that $d(f(x), g(x)) < \delta \Rightarrow$ both $(f(x), g(x)) \in M$ and

$$d(f(x),\ a(f(x), g(x),\ t)) < \varepsilon,\ \forall t \in I.$$

Now determine the order l such that

$$(2\,n+1) \text{ mesh } L_0 < \delta.$$

On setting $g_t(x) = a(f(x), g(x),\ t)$,we obtain the conclusion (ii).

□

By means of Theorem 2.5, we can prove the following without any difficulty:

2.9 Theorem. *Let L be a connected finite simplicial complex, U an open subset of* $|L|$, *and* $f: U \to |L|$ *a mapping. If the fixed point set* $\Phi(f)$ *of* f *is compact, then for any given number* $\varepsilon > 0$, *there exists a homotopy* $f_t: U \to |L|$, $t \in I$, *such that*

(i) $f_0 = f$;
(ii) $d(f_t(x), f(x)) < \varepsilon$, $\forall (x, t) \in U \times I$;
(iii) $f_t(x) = f(x)$, $\forall x \notin$ *certain compact subset of* U, $t \in I$;
(iv) f_1 *has only a finite number of fixed points, all are in the interiors of maximal simplexes of* L, *and* f_1 *is linear in a neighborhood of each fixed point.*

proof. Take a neighborhood V of $\Phi(f)$ such that $\bar{V} \subseteq U$. Take a subcomplex K of a subdivision $\mathrm{Sd}^{(l)} L$ of sufficiently high order l such that $\bar{V} \subseteq \mathrm{Int}\, |K| \subseteq |K| \subseteq U$. Since f has no fixed point on the compact set $|K| - V$, there exists a number $\eta > 0$, but $< \varepsilon$, such that

$$d(x, f(x)) \geqslant \eta, \quad x \in |K| - V.$$

By virtue of Theorem 2.5, there exists a homotopy $g_t: |K| \to |L|$, $t \in I$, such that $g_0 = f$, $d(g_t(x), f(x)) < \eta$, and g_1 has only a finite number of fixed points in the interiors of maximal simplexes of L, and is linear in a neighborhood of each fixed point.

Denote the boundary of the subspace $|K|$ in the polyhedron $|L|$ by $\partial |K|$. Obviously $\partial |K| \cap \bar{V} = \emptyset$. Let $\lambda: |K| \to I$ be such that ([8] Theorem II 6.3)

$$\lambda(x) = \begin{cases} 0, & \text{when } x \in \partial |K|, \\ 1, & \text{when } x \in \bar{V}. \end{cases}$$

It is easy to see that $f_t: U \to |L|$, $t \in I$, defined by

$$f_t(x) = \begin{cases} f(x), & \text{when } x \in U - |K|, \\ g_{t\lambda(x)}(x), & \text{when } x \in |K|, \end{cases}$$

is a homotopy, which has the first three properties. Moreover,

$$d(x, f(x)) \geqslant \eta > d(g_{\lambda(x)}(x), f(x)) = d(f_1(x), f(x)), \quad x \in |K| - V.$$

Hence the fixed points of f_1 are all on V and $f_1 = g_1$ on V as well; in other words, the homotopy has also the property (iv). □

Bibliography

Books

[1] Alexandroff, P. & Hopf, H., *Topologie*, Springer, Berlin, 1935.

[2] Brown, R. F., *The Lefschetz Fixed Point Theorem*, Scott, Foresman and Company, Glenview, Illinois, 1971.

[2*] Bibliography of recent papers on fixed point theory, *Rocky Mountain J. Math.*, **4** (1974), 107—134.

[3] Croom, F. H., *Basic Concepts of Algebraic Topology*, Springer, 1978.

[4] Cronin, J., *Fixed Points and Topological Degree in Nonlinear Analysis*, Amer. Math. Soc., 1964.

[5] Dold, A., *Lectures on Algebraic Topology*, Springer, 1972.

[6] Dugundji, J. & Granas, A., *Fixed Point Theory*, Vol. 1, Polish Scientific Publishers, Warsaw, 1982.

[7] Eisenack,G. & Fenske, C., *Fixpunkttheorie*, Bibliographischer Institut, Mannheim, Wien, Zürich, 1978.

[8] Kiang, T. H. *Introduction to Algebraic Topology* (in Chinese), Shanghai Sci. and Tech. Press, 1978.

[9] Lefschetz, S., *Introduction to Topology*, Princeton, 1949.

[10] Massey, W. S., *Algebraic Topology, an Introduction*, Harcourt, Brace & World, Inc., 1967.

[11] Seifert, H. & Threlfall,W., *Lehrbuch der Topologie*, Leipzig, 1934. English version: *A Textbook of Topology*, by Goldman, M. A., Academic Press, 1980.

[12] Singer, I. M. & Thorpe, J. A., *Lecture Notes on Elementary Topology and Geometry*, Scott, Foresman and Company, Glenview, Illinois, 1967.

[13] Smart, D. R., *Fixed Point Theorems*, Cambridge University Press, 1974.

[14] van der Walt, T., *Fixed and Almost Fixed Points*, Amsterdam, 1967.

Papers

[15] Amann,H. & Weiss S. A., On the uniqueness of the topological degree, *Math. Z.*, **130** (1973), 39—54.

[16] Barnier, W., The Jiang subgroup for a map, Doctorial Dissertation, University of California, Los Angeles, 1967.

[17] Brooks, R., Certain subgroups of the fundamental group and the number of roots of $f(x)=a$, *Amer. J. Math.*, **95** (1973), 720—728.

[18] Brown, R. F., On some old problems of fixed point theory, *Rocky Mountain J. Math.*, **4** (1974), 3—14.

[19] Chang, H. C., On B. J. Jiang's three lemmas, *The Advances in Math.* (in Chinese), **11** (1982), 77—80.

[20] Fadell, E., Recent results in the fixed point theory of continuous maps, *Bull. Amer. Math. Soc.*, **76** (1970), 10—29.

[21] Fenchel, W., Jacob Nielsen in memoriam, *Acta Math.*, **103** (1960): 3—4, vii—xix.

[22] Franz, W., Abbildungsklassen und Fixpunktklassen dreidimensionaler Linsenräume, *Crelle J.*, **185** (1943), 65—77.

[23] Gottlieb,D., A certain subgroup of the fundamental group, *Amer. J. Math.*, **87** (1965), 840—856.

[24] Hopf, H., A new proof of the Lefschetz formula on invariant points, *Proc. Nat. Acad. Sci., U. S. A.*, **14** (1928), 149—153.

[25] Jiang, B. J., Estimation of the Nielsen numbers, I, *Acta Math. Sinica* (in Chinese), **14** (1964), 304—312, or its English version, *Chinese Math.-Acta,* **5** (1964), 330—339; II, *Acta Sci. Natur. Univ. Peking* (in Chinese), 1979, 48—57.

[26] Kiang, T. H., On the groups of orientable two-manifolds, *Proc. Nat. Acad, Sci., U. S. A.,* **17** (1931), 142—144.

[26 a] ——, On the Poincarés groups and the extended universal coverings of closed orientable two-manifolds, *Jour. Chinese Math. Soc.,* I (1936), 43—153.

[26 b] ——, Recent developement of the theory of fixed point classes, *Applied and Computational Math.* (in Chinese), I (1979), 53—59.

[27] Kiang, T. H. & Jiang B. J., The Nielsen numbers of selfmappings of the same homotopy type, *Sci. Sinica,* **12** (1963), 1071—1072.

[28] Lefchetz, S., Continuous transformations of manifolds, *Proc. Nat. Acad. Sci., U. S. A.,* **9** (1923), 90—93.

[29] Leray, J. La théorie des points fixes et ses applications en analyse, *Proc. Internat. Congr. Math.,* Cambridge, 1950, Vol. II, 202—208.

[30] McCord, D., An estimate of the Nielsen number and an example concerning the Lefschetz fixed point theorem, *Pacific J. Math.,* **66** (1976), 195—203.

[31] Newman, M. H. A., Fixed point and coincidence theorems, *J. London Math. Soc.,* **27** (1952), 135—140.

[32] Nielsen, J., Untersuchungen zur Topologie der geschlossenen zweiseitigen Flächen, I, II, III, *Acta Math.,* **50** (1927), 189—358; **53** (1929), 1—76; **58** (1932), 87—167; IV, *Det. Kgl. Danske Videnskabernes Salskab, Math. - fis. Medd.,* Band XXI/2, 1944.

[33] Olum, P., Mappings of manifolds and the notion of degree, *Annals of Math.,* **58** (1953), 458—480.

[34] Scholz, K. U., The Nielsen fixed point theory for noncompact spaces, *Rocky Mountain J. Math.,* **4** (1974), 81—87.

[35] Shi, G. H., On the least number of fixed points and the Nielsen numbers, *Acta Math. Sinica* (in Chinese), **16** (1966), 223—232, or its English version, *Chinese Math.-Acta,* **8** (1966), 234—243.

[35 a] ——, The least number of fixed points of the identity mapping class, *Acta Math. Sinica* (in Chinese), **18** (1975), 192—202.

[36] Wecken, F., Fixpunktklassen, I, II, III, *Math. Ann.,* **117** (1941), 659—671; **118** (1942), 216—234, 544—577.

[37] Zeidler, E., Existenz, Eindeutigkeit, Eigenschaften und Anwendungen des Abbildungsgrades im R^n, Theory of Nonlinear Operators, Proc. of a summer school held in Oct., 1972 at Neuendorf, GDR, 1974, 259—312.

Epilogue

In the annual meeting of Peking Mathematical Society, held in August of 1964, I presented at the Section of Geometry and Topology the paper "The theory of fixed point classes and some of its new advances". I planned at that time to write as a continuation of [8] a treatise as outlined in this paper. This became almost untenable during the period of "The Gang of Four". However, [20] was accessible to me in about 1972, and a copy of [2] was sent to me as a gift from a friend in the April of 1973. They made me recall my plan of almost ten years old. It turned out in the spring of 1975 that I had my enthusiasm revived and also a little time for returning to my old plan. As I felt I was getting on in years, I had to make best use of my time. I started to write the present treatise, summarizing mainly our own work and presenting the theory in a more elementary way as we had looked at it. I worked alone and silently for almost two years, and finished only a first draft of the first three chapters and of the first two appendices of the treatise. Realizing that it would be impossible for a publisher to print such a mathematical treatise at that time, I had 100 copies of this first draft printed in a very primitive way in the November of 1976 with the aids from several friends. Under the title on the front page of this draft copy there was printed the entry "(1927—1976)" to commemorate the half century anniversary of [32,I].

In the April of 1977, that is, about half a year after the downfall of "The Gang of Four", The Science Press, Academia Sinica, took the initiative step and consented to print my treatise when completed. In order to have the treatise completed sooner, our department made the arrangement that at the beginning from November of that year, four of my former students gathered in our department and worked together with me on it. They are Messrs. Jiang Boju of our department, Shi Genhua from the Ministry of Water Conservancy and Electric Power, Ding Zhengliang from the University of Science and Technology of China, Hefei, and You Chengye of our department.

Jiang prepared a draft of II§ 4, II§ 7 and C§ 2, Shi that of IV and C§ 1, while Ding and I that of V. For four months we five met regularly; wrote, discussed, and made revisions repeatedly. The final manuscript was completed and accepted for publication by The Science Press at the end of 1977. Now during the proof-reading (I am helped by my former student Liu Yingming of Sichuan University, and colleagues Li Tongfu and You Chengye of our department), I have the opportunity to add this epilogue and to acknowledge that this treatise is in fact a work of cooperation. To The Science Press, to our department, and to every of the friends mentioned above, I am heartily grateful.

Kiang Tsai-han

Peking University
April 1979

List of symbols

The symbol □ signifies the end of a proof, or the end of an easy proposition without proof. Theorem 2.1, Theorem III 2.1, or Theorem C 2.1 in a sentence means the Theorem 2.1 in the same chapter or appendix, in Chapter III, or in Appendix C, respectively.

Index